NEW CAMBRIDGE
STATISTICAL
TABLES

D. V. LINDLEY
&
W. F. SCOTT

Second Edition

CAMBRIDGE
UNIVERSITY PRESS

CONTENTS

CONVENTION. To prevent the tables becoming too dense with figures, the convention has been adopted of omitting the leading figure when this does not change too often, only including it at the beginning of a set of five entries, or when it changes. (Table 23 provides an example.)

PREFACE TO THE FIRST EDITION

The raison d'être of this set of tables is the same as that of the set it replaces, the *Cambridge Elementary Statistical Tables* (Lindley and Miller, 1953), and is described in the first paragraph of their preface.

This set of tables is concerned only with the commoner and more familiar and elementary of the many statistical functions and tests of significance now available. It is hoped that the values provided will meet the majority of the needs of many users of statistical methods in scientific research, technology and industry in a compact and handy form, and that the collection will provide a convenient set of tables for the teaching and study of statistics in schools and universities.

The concept of what constitutes a familiar or elementary statistical procedure has changed in 30 years and, as a result, many statistical tables not in the earlier set have been included, together with tables of the binomial, hypergeometric and Poisson distributions. A large part of the earlier set of tables consisted of functions of the integers. These are now readily available elsewhere, or can be found using even the simplest of pocket calculators, and have therefore been omitted.

The binomial, Poisson, hypergeometric, normal, χ^2 and t distributions have been fully tabulated so that all values within the ranges of the arguments chosen can be found. Linear, and in some cases quadratic or harmonic, interpolation will sometimes be necessary and a note on this has been provided. Most of the other tables give only the percentage points of distributions, sufficient to carry out significance tests at the usual 5 per cent and 1 per cent levels, both one- and two-sided, and there are also some 10 per cent, 2·5 per cent and 0·1 per cent points. Limitation of space has forced the number of levels to be reduced in some cases. Besides distributions, there are tables of binomial coefficients, random sampling numbers, random normal deviates and logarithms of factorials.

Each table is accompanied by a brief description of what is tabulated and, where the table is for a specific usage, a description of that is given. With the exception of Table 26, no attempt has been made to provide accounts of other statistical procedures that use the tables or to illustrate their use with numerical examples, it being felt that these are more appropriate in an accompanying text or otherwise provided by the teacher.

The choice of which tables to include has been influenced by the student's need to follow prescribed syllabuses and to pass the associated examinations. The inclusion of a table does not therefore imply the authors' endorsement of the technique associated with it. This is true of some significance tests, which could be more informatively replaced by robust estimates of the parameter being tested, together with a standard error.

All significance tests are dubious because the interpretation to be placed on the phrase 'significant at 5%' depends on the sample size: it is more indicative of the falsity of the null hypothesis with a small sample than with a large one. In addition, any test of the hypothesis that a parameter takes a specified value is dubious because significance at a prescribed level can generally be achieved by taking a large enough sample (cf. M. H. DeGroot, *Probability and Statistics* (1975), Addison-Wesley, p. 421).

All the values here are exact to the number of places given, except that in Table 14 the values for $n > 17$ were calculated by an Edgeworth series approximation described in 'Critical values of the coefficient of rank correlation for testing the hypothesis of independence' by G. J. Glasser and R. F. Winter, *Biometrika* **48** (1961), pp. 444–8.

Nearly all the tables have been newly computed for this publication and compared with existing compilations: the exceptions, in which we have used material from other sources, are listed below:

Table 14, $n = 12$ to 16, is taken from 'The null distribution of Spearman's S when $n = 13(1)16$', by A. Otten, *Statistica Neerlandica*, **27** (1973), pp. 19–20, by permission of the editor.

Table 24, $k = 6$, $n = 5$ and 6, is taken from 'Extended tables of the distribution of Friedman's S-statistic in the two-way layout', by Robert E. Odeh, *Commun. Statist. – Simula Computa.*, **B6** (1), 29–48 (1977), by permission of Marcel Dekker, Inc., and from Table 39 of *The Pocket Book of Statistical Tables*, by Robert E. Odeh, Donald B. Owen, Z. W. Birnbaum and Lloyd Fisher, Marcel Dekker (1977), by permission of Marcel Dekker, Inc.

Table 25, $k = 3$, 4, 5, is partly taken from 'Exact probability levels for the Kruskal–Wallis test', by Ronald L. Iman, Dana Quade and Douglas A. Alexander, *Selected Tables in Mathematical Statistics*, Vol. 3 (1975), by permission of the American Mathematical Society; $k = 3$ is also partly taken from the MS thesis of Douglas A. Alexander, University of North Carolina at Chapel Hill (1968), by permission of Douglas A. Alexander.

We should like to thank the staff of the University Press for their helpful advice and co-operation during the printing of the tables. We should also like to thank the staff of Heriot-Watt University's Computer Centre and Mr Ian Sweeney for help with some computing aspects.

10 January 1984

PREFACE TO THE SECOND EDITION

The only change from the first edition is the inclusion of tables of Bayesian confidence intervals for the binomial and Poisson distributions and for the square of a multiple correlation coefficient.

D. V. Lindley
2 Periton Lane,
Minehead
Somerset,
TA24 8AQ, U.K.

W. F. Scott
Department of Actuarial Mathematics
and Statistics,
Heriot-Watt University
Riccarton, Edinburgh
EH14 4AS, U.K.

TABLE 1. THE BINOMIAL DISTRIBUTION FUNCTION

$n=2$ $r=0$	1		$n=3$ $r=0$	1	2
$p=0.01$ 0.9801	0.9999		$p=0.01$ 0.9703	0.9997	
.02 .9604	.9996		.02 .9412	.9988	
.03 .9409	.9991		.03 .9127	.9974	
.04 .9216	.9984		.04 .8847	.9953	0.9999
0.05 0.9025	0.9975		0.05 0.8574	0.9928	0.9999
.06 .8836	.9964		.06 .8306	.9896	.9998
.07 .8649	.9951		.07 .8044	.9860	.9997
.08 .8464	.9936		.08 .7787	.9818	.9995
.09 .8281	.9919		.09 .7536	.9772	.9993
0.10 0.8100	0.9900		0.10 0.7290	0.9720	0.9990
.11 .7921	.9879		.11 .7050	.9664	.9987
.12 .7744	.9856		.12 .6815	.9603	.9983
.13 .7569	.9831		.13 .6585	.9537	.9978
.14 .7396	.9804		.14 .6361	.9467	.9973
0.15 0.7225	0.9775		0.15 0.6141	0.9393	0.9966
.16 .7056	.9744		.16 .5927	.9314	.9959
.17 .6889	.9711		.17 .5718	.9231	.9951
.18 .6724	.9676		.18 .5514	.9145	.9942
.19 .6561	.9639		.19 .5314	.9054	.9931
0.20 0.6400	0.9600		0.20 0.5120	0.8960	0.9920
.21 .6241	.9559		.21 .4930	.8862	.9907
.22 .6084	.9516		.22 .4746	.8761	.9894
.23 .5929	.9471		.23 .4565	.8656	.9878
.24 .5776	.9424		.24 .4390	.8548	.9862
0.25 .5625	0.9375		0.25 0.4219	0.8438	0.9844
.26 .5476	.9324		.26 .4052	.8324	.9824
.27 .5329	.9271		.27 .3890	.8207	.9803
.28 .5184	.9216		.28 .3732	.8087	.9780
.29 .5041	.9159		.29 .3579	.7965	.9756
0.30 0.4900	0.9100		0.30 0.3430	0.7840	0.9730
.31 .4761	.9039		.31 .3285	.7713	.9702
.32 .4624	.8976		.32 .3144	.7583	.9672
.33 .4489	.8911		.33 .3008	.7452	.9641
.34 .4356	.8844		.34 .2875	.7318	.9607
0.35 0.4225	0.8775		0.35 0.2746	0.7182	0.9571
.36 .4096	.8704		.36 .2621	.7045	.9533
.37 .3969	.8631		.37 .2500	.6906	.9493
.38 .3844	.8556		.38 .2383	.6765	.9451
.39 .3721	.8479		.39 .2270	.6623	.9407
0.40 0.3600	0.8400		0.40 0.2160	0.6480	0.9360
.41 .3481	.8319		.41 .2054	.6335	.9311
.42 .3364	.8236		.42 .1951	.6190	.9259
.43 .3249	.8151		.43 .1852	.6043	.9205
.44 .3136	.8064		.44 .1756	.5896	.9148
0.45 0.3025	0.7975		0.45 0.1664	0.5748	0.9089
.46 .2916	.7884		.46 .1575	.5599	.9027
.47 .2809	.7791		.47 .1489	.5449	.8962
.48 .2704	.7696		.48 .1406	.5300	.8894
.49 .2601	.7599		.49 .1327	.5150	.8824
0.50 0.2500	0.7500		0.50 0.1250	0.5000	0.8750

The function tabulated is

$$F(r|n,p) = \sum_{t=0}^{r} \binom{n}{t} p^t (1-p)^{n-t}$$

for $r = 0, 1, \ldots, n-1$, $n \leq 20$ and $p \leq 0.5$; n is sometimes referred to as the index and p as the parameter of the distribution. $F(r|n, p)$ is the probability that X, the number of occurrences in n independent trials of an event with probability p of occurrence in each trial, is less than or equal to r; that is,

$$\Pr\{X \leq r\} = F(r|n,p).$$

Note that

$$\Pr\{X \geq r\} = 1 - \Pr\{X \leq r-1\}$$
$$= 1 - F(r-1|n,p).$$

$F(n|n, p) = 1$, and the values for $p > 0.5$ may be found using the result

$$F(r|n,p) = 1 - F(n-r-1|n, 1-p).$$

The probability of *exactly* r occurrences, $\Pr\{X = r\}$, is equal to

$$F(r|n, p) - F(r-1|n, p) = \binom{n}{r} p^r (1-p)^{n-r}.$$

Linear interpolation in p is satisfactory over much of the table but there are places where quadratic interpolation is necessary for high accuracy. When $r = 0, 1$ or $n-1$ a direct calculation is to be preferred:

$$F(0|n, p) = (1-p)^n,$$
$$F(1|n, p) = (1-p)^{n-1}[1 + (n-1)p]$$

and

$$F(n-1|n, p) = 1 - p^n.$$

For $n > 20$ the number of occurrences X is approximately normally distributed with mean np and variance $np(1-p)$; hence, including $\frac{1}{2}$ for continuity, we have

$$F(r|n,p) \doteq \Phi(s)$$

where $s = \dfrac{r + \frac{1}{2} - np}{\sqrt{np(1-p)}}$ and $\Phi(s)$ is the normal distribution function (see Table 4). The approximation can usually be improved by using the formula

$$F(r|n,p) \doteq \Phi(s) - \frac{\gamma}{6\sqrt{2\pi}} e^{-\frac{1}{2}s^2}(s^2 - 1)$$

where $\gamma = \dfrac{1-2p}{\sqrt{np(1-p)}}$.

An alternative approximation for $n > 20$ when p is small and np is of moderate size is to use the Poisson distribution:

$$F(r|n,p) \doteq F(r|\mu)$$

where $\mu = np$ and $F(r|\mu)$ is the Poisson distribution function (see Table 2). If $1-p$ is small and $n(1-p)$ is of moderate size a similar approximation gives

$$F(r|n,p) \doteq 1 - F(n-r-1|\mu)$$

where $\mu = n(1-p)$.

Omitted entries to the left and right of tabulated values are 0 and 1 respectively, to four decimal places.

TABLE 1. THE BINOMIAL DISTRIBUTION FUNCTION

$n = 4$	$r = 0$	1	2	3	$n = 5$	$r = 0$	1	2	3	4
$p = 0{\cdot}01$	0·9606	0·9994			$p = 0{\cdot}01$	0·9510	0·9990			
·02	·9224	·9977			·02	·9039	·9962	0·9999		
·03	·8853	·9948	0·9999		·03	·8587	·9915	·9997		
·04	·8493	·9909	·9998		·04	·8154	·9852	·9994		
0·05	0·8145	0·9860	0·9995		0·05	0·7738	0·9774	0·9988		
·06	·7807	·9801	·9992		·06	·7339	·9681	·9980	0·9999	
·07	·7481	·9733	·9987		·07	·6957	·9575	·9969	·9999	
·08	·7164	·9656	·9981		·08	·6591	·9456	·9955	·9998	
·09	·6857	·9570	·9973	0·9999	·09	·6240	·9326	·9937	·9997	
0·10	0·6561	0·9477	0·9963	0·9999	0·10	0·5905	0·9185	0·9914	0·9995	
·11	·6274	·9376	·9951	·9999	·11	·5584	·9035	·9888	·9993	
·12	·5997	·9268	·9937	·9998	·12	·5277	·8875	·9857	·9991	
·13	·5729	·9153	·9921	·9997	·13	·4984	·8708	·9821	·9987	
·14	·5470	·9032	·9902	·9996	·14	·4704	·8533	·9780	·9983	0·9999
0·15	0·5220	0·8905	0·9880	0·9995	0·15	0·4437	0·8352	0·9734	0·9978	0·9999
·16	·4979	·8772	·9856	·9993	·16	·4182	·8165	·9682	·9971	·9999
·17	·4746	·8634	·9829	·9992	·17	·3939	·7973	·9625	·9964	·9999
·18	·4521	·8491	·9798	·9990	·18	·3707	·7776	·9563	·9955	·9998
·19	·4305	·8344	·9765	·9987	·19	·3487	·7576	·9495	·9945	·9998
0·20	0·4096	0·8192	0·9728	0·9984	0·20	0·3277	0·7373	0·9421	0·9933	0·9997
·21	·3895	·8037	·9688	·9981	·21	·3077	·7167	·9341	·9919	·9996
·22	·3702	·7878	·9644	·9977	·22	·2887	·6959	·9256	·9903	·9995
·23	·3515	·7715	·9597	·9972	·23	·2707	·6749	·9164	·9886	·9994
·24	·3336	·7550	·9547	·9967	·24	·2536	·6539	·9067	·9866	·9992
0·25	0·3164	0·7383	0·9492	0·9961	0·25	0·2373	0·6328	0·8965	0·9844	0·9990
·26	·2999	·7213	·9434	·9954	·26	·2219	·6117	·8857	·9819	·9988
·27	·2840	·7041	·9372	·9947	·27	·2073	·5907	·8743	·9792	·9986
·28	·2687	·6868	·9306	·9939	·28	·1935	·5697	·8624	·9762	·9983
·29	·2541	·6693	·9237	·9929	·29	·1804	·5489	·8499	·9728	·9979
0·30	0·2401	0·6517	0·9163	0·9919	0·30	0·1681	0·5282	0·8369	0·9692	0·9976
·31	·2267	·6340	·9085	·9908	·31	·1564	·5077	·8234	·9653	·9971
·32	·2138	·6163	·9004	·9895	·32	·1454	·4875	·8095	·9610	·9966
·33	·2015	·5985	·8918	·9881	·33	·1350	·4675	·7950	·9564	·9961
·34	·1897	·5807	·8829	·9866	·34	·1252	·4478	·7801	·9514	·9955
0·35	0·1785	0·5630	0·8735	0·9850	0·35	0·1160	0·4284	0·7648	0·9460	0·9947
·36	·1678	·5453	·8638	·9832	·36	·1074	·4094	·7491	·9402	·9940
·37	·1575	·5276	·8536	·9813	·37	·0992	·3907	·7330	·9340	·9931
·38	·1478	·5100	·8431	·9791	·38	·0916	·3724	·7165	·9274	·9921
·39	·1385	·4925	·8321	·9769	·39	·0845	·3545	·6997	·9204	·9910
0·40	0·1296	0·4752	0·8208	0·9744	0·40	0·0778	0·3370	0·6826	0·9130	0·9898
·41	·1212	·4580	·8091	·9717	·41	·0715	·3199	·6651	·9051	·9884
·42	·1132	·4410	·7970	·9689	·42	·0656	·3033	·6475	·8967	·9869
·43	·1056	·4241	·7845	·9658	·43	·0602	·2871	·6295	·8879	·9853
·44	·0983	·4074	·7717	·9625	·44	·0551	·2714	·6114	·8786	·9835
0·45	0·0915	0·3910	0·7585	0·9590	0·45	0·0503	0·2562	0·5931	0·8688	0·9815
·46	·0850	·3748	·7450	·9552	·46	·0459	·2415	·5747	·8585	·9794
·47	·0789	·3588	·7311	·9512	·47	·0418	·2272	·5561	·8478	·9771
·48	·0731	·3431	·7169	·9469	·48	·0380	·2135	·5375	·8365	·9745
·49	·0677	·3276	·7023	·9424	·49	·0345	·2002	·5187	·8248	·9718
0·50	0·0625	0·3125	0·6875	0·9375	0·50	0·0313	0·1875	0·5000	0·8125	0·9688

See page 4 for explanation of the use of this table.

TABLE 1. THE BINOMIAL DISTRIBUTION FUNCTION

$n = 6$

p	$r=0$	1	2	3	4	5
0·01	0·9415	0·9985				
·02	·8858	·9943	0·9998			
·03	·8330	·9875	·9995			
·04	·7828	·9784	·9988			
0·05	0·7351	0·9672	0·9978	0·9999		
·06	·6899	·9541	·9962	·9998		
·07	·6470	·9392	·9942	·9997		
·08	·6064	·9227	·9915	·9995		
·09	·5679	·9048	·9882	·9992		
0·10	0·5314	0·8857	0·9842	0·9987	0·9999	
·11	·4970	·8655	·9794	·9982	·9999	
·12	·4644	·8444	·9739	·9975	·9999	
·13	·4336	·8224	·9676	·9966	·9998	
·14	·4046	·7997	·9605	·9955	·9997	
0·15	0·3771	0·7765	0·9527	0·9941	0·9996	
·16	·3513	·7528	·9440	·9925	·9995	
·17	·3269	·7287	·9345	·9906	·9993	
·18	·3040	·7044	·9241	·9884	·9990	
·19	·2824	·6799	·9130	·9859	·9987	
0·20	0·2621	0·6554	0·9011	0·9830	0·9984	0·9999
·21	·2431	·6308	·8885	·9798	·9980	·9999
·22	·2252	·6063	·8750	·9761	·9975	·9999
·23	·2084	·5820	·8609	·9720	·9969	·9999
·24	·1927	·5578	·8461	·9674	·9962	·9998
0·25	0·1780	0·5339	0·8306	0·9624	0·9954	0·9998
·26	·1642	·5104	·8144	·9569	·9944	·9997
·27	·1513	·4872	·7977	·9508	·9933	·9996
·28	·1393	·4644	·7804	·9443	·9921	·9995
·29	·1281	·4420	·7626	·9372	·9907	·9994
0·30	0·1176	0·4202	0·7443	0·9295	0·9891	0·9993
·31	·1079	·3988	·7256	·9213	·9873	·9991
·32	·0989	·3780	·7064	·9125	·9852	·9989
·33	·0905	·3578	·6870	·9031	·9830	·9987
·34	·0827	·3381	·6672	·8931	·9805	·9985
0·35	0·0754	0·3191	0·6471	0·8826	0·9777	0·9982
·36	·0687	·3006	·6268	·8714	·9746	·9978
·37	·0625	·2828	·6063	·8596	·9712	·9974
·38	·0568	·2657	·5857	·8473	·9675	·9970
·39	·0515	·2492	·5650	·8343	·9635	·9965
0·40	0·0467	0·2333	0·5443	0·8208	0·9590	0·9959
·41	·0422	·2181	·5236	·8067	·9542	·9952
·42	·0381	·2035	·5029	·7920	·9490	·9945
·43	·0343	·1895	·4823	·7768	·9434	·9937
·44	·0308	·1762	·4618	·7610	·9373	·9927
0·45	0·0277	0·1636	0·4415	0·7447	0·9308	0·9917
·46	·0248	·1515	·4214	·7279	·9238	·9905
·47	·0222	·1401	·4015	·7107	·9163	·9892
·48	·0198	·1293	·3820	·6930	·9083	·9878
·49	·0176	·1190	·3627	·6748	·8997	·9862
0·50	0·0156	0·1094	0·3438	0·6562	0·8906	0·9844

$n = 7$

p	$r=0$	1	2	3
0·01	0·9321	0·9980		
·02	·8681	·9921	0·9997	
·03	·8080	·9829	·9991	
·04	·7514	·9706	·9980	0·9999
0·05	0·6983	0·9556	0·9962	0·9998
·06	·6485	·9382	·9937	·9996
·07	·6017	·9187	·9903	·9993
·08	·5578	·8974	·9860	·9988
·09	·5168	·8745	·9807	·9982
0·10	0·4783	0·8503	0·9743	0·9973
·11	·4423	·8250	·9669	·9961
·12	·4087	·7988	·9584	·9946
·13	·3773	·7719	·9487	·9928
·14	·3479	·7444	·9380	·9906
0·15	0·3206	0·7166	0·9262	0·9879
·16	·2951	·6885	·9134	·9847
·17	·2714	·6604	·8995	·9811
·18	·2493	·6323	·8846	·9769
·19	·2288	·6044	·8687	·9721
0·20	0·2097	0·5767	0·8520	0·9667
·21	·1920	·5494	·8343	·9606
·22	·1757	·5225	·8159	·9539
·23	·1605	·4960	·7967	·9464
·24	·1465	·4702	·7769	·9383
0·25	0·1335	0·4449	0·7564	0·9294
·26	·1215	·4204	·7354	·9198
·27	·1105	·3965	·7139	·9095
·28	·1003	·3734	·6919	·8984
·29	·0910	·3510	·6696	·8866
0·30	0·0824	0·3294	0·6471	0·8740
·31	·0745	·3086	·6243	·8606
·32	·0672	·2887	·6013	·8466
·33	·0606	·2696	·5783	·8318
·34	·0546	·2513	·5553	·8163
0·35	0·0490	0·2338	0·5323	0·8002
·36	·0440	·2172	·5094	·7833
·37	·0394	·2013	·4866	·7659
·38	·0352	·1863	·4641	·7479
·39	·0314	·1721	·4419	·7293
0·40	0·0280	0·1586	0·4199	0·7102
·41	·0249	·1459	·3983	·6906
·42	·0221	·1340	·3771	·6706
·43	·0195	·1228	·3564	·6502
·44	·0173	·1123	·3362	·6294
0·45	0·0152	0·1024	0·3164	0·6083
·46	·0134	·0932	·2973	·5869
·47	·0117	·0847	·2787	·5654
·48	·0103	·0767	·2607	·5437
·49	·0090	·0693	·2433	·5219
0·50	0·0078	0·0625	0·2266	0·5000

See page 4 for explanation of the use of this table.

TABLE 1. THE BINOMIAL DISTRIBUTION FUNCTION

$n = 7$	$r = 4$	5	6
$p = 0.01$			
.02			
.03			
.04			
0.05			
.06			
.07			
.08	0.9999		
.09	.9999		
0.10	0.9998		
.11	.9997		
.12	.9996		
.13	.9994		
.14	.9991		
0.15	0.9988	0.9999	
.16	.9983	.9999	
.17	.9978	.9999	
.18	.9971	.9998	
.19	.9963	.9997	
0.20	0.9953	0.9996	
.21	.9942	.9995	
.22	.9928	.9994	
.23	.9912	.9992	
.24	.9893	.9989	
0.25	0.9871	0.9987	0.9999
.26	.9847	.9983	.9999
.27	.9819	.9979	.9999
.28	.9787	.9974	.9999
.29	.9752	.9969	.9998
0.30	0.9712	0.9962	0.9998
.31	.9668	.9954	.9997
.32	.9620	.9945	.9997
.33	.9566	.9935	.9996
.34	.9508	.9923	.9995
0.35	0.9444	0.9910	0.9994
.36	.9375	.9895	.9992
.37	.9299	.9877	.9991
.38	.9218	.9858	.9989
.39	.9131	.9836	.9986
0.40	0.9037	0.9812	0.9984
.41	.8937	.9784	.9981
.42	.8831	.9754	.9977
.43	.8718	.9721	.9973
.44	.8598	.9684	.9968
0.45	0.8471	0.9643	0.9963
.46	.8337	.9598	.9956
.47	.8197	.9549	.9949
.48	.8049	.9496	.9941
.49	.7895	.9438	.9932
0.50	0.7734	0.9375	0.9922

$n = 8$	$r = 0$	1	2	3	4	5	6
$p = 0.01$	0.9227	0.9973	0.9999				
.02	.8508	.9897	.9996				
.03	.7837	.9777	.9987	0.9999			
.04	.7214	.9619	.9969	.9998			
0.05	0.6634	0.9428	0.9942	0.9996			
.06	.6096	.9208	.9904	.9993			
.07	.5596	.8965	.9853	.9987	0.9999		
.08	.5132	.8702	.9789	.9978	.9999		
.09	.4703	.8423	.9711	.9966	.9997		
0.10	0.4305	0.8131	0.9619	0.9950	0.9996		
.11	.3937	.7829	.9513	.9929	.9993		
.12	.3596	.7520	.9392	.9903	.9990	0.9999	
.13	.3282	.7206	.9257	.9871	.9985	.9999	
.14	.2992	.6889	.9109	.9832	.9979	.9998	
0.15	0.2725	0.6572	0.8948	0.9786	0.9971	0.9998	
.16	.2479	.6256	.8774	.9733	.9962	.9997	
.17	.2252	.5943	.8588	.9672	.9950	.9995	
.18	.2044	.5634	.8392	.9603	.9935	.9993	
.19	.1853	.5330	.8185	.9524	.9917	.9991	0.9999
0.20	0.1678	0.5033	0.7969	0.9437	0.9896	0.9988	0.9999
.21	.1517	.4743	.7745	.9341	.9871	.9984	.9999
.22	.1370	.4462	.7514	.9235	.9842	.9979	.9998
.23	.1236	.4189	.7276	.9120	.9809	.9973	.9998
.24	.1113	.3925	.7033	.8996	.9770	.9966	.9997
0.25	0.1001	0.3671	0.6785	0.8862	0.9727	0.9958	0.9996
.26	.0899	.3427	.6535	.8719	.9678	.9948	.9995
.27	.0806	.3193	.6282	.8567	.9623	.9936	.9994
.28	.0722	.2969	.6027	.8406	.9562	.9922	.9992
.29	.0646	.2756	.5772	.8237	.9495	.9906	.9990
0.30	0.0576	0.2553	0.5518	0.8059	0.9420	0.9887	0.9987
.31	.0514	.2360	.5264	.7874	.9339	.9866	.9984
.32	.0457	.2178	.5013	.7681	.9250	.9841	.9980
.33	.0406	.2006	.4764	.7481	.9154	.9813	.9976
.34	.0360	.1844	.4519	.7276	.9051	.9782	.9970
0.35	0.0319	0.1691	0.4278	0.7064	0.8939	0.9747	0.9964
.36	.0281	.1548	.4042	.6847	.8820	.9707	.9957
.37	.0248	.1414	.3811	.6626	.8693	.9664	.9949
.38	.0218	.1289	.3585	.6401	.8557	.9615	.9939
.39	.0192	.1172	.3366	.6172	.8414	.9561	.9928
0.40	0.0168	0.1064	0.3154	0.5941	0.8263	0.9502	0.9915
.41	.0147	.0963	.2948	.5708	.8105	.9437	.9900
.42	.0128	.0870	.2750	.5473	.7938	.9366	.9883
.43	.0111	.0784	.2560	.5238	.7765	.9289	.9864
.44	.0097	.0705	.2376	.5004	.7584	.9206	.9843
0.45	0.0084	0.0632	0.2201	0.4770	0.7396	0.9115	0.9819
.46	.0072	.0565	.2034	.4537	.7202	.9018	.9792
.47	.0062	.0504	.1875	.4306	.7001	.8914	.9761
.48	.0053	.0448	.1724	.4078	.6795	.8802	.9728
.49	.0046	.0398	.1581	.3854	.6584	.8682	.9690
0.50	0.0039	0.0352	0.1445	0.3633	0.6367	0.8555	0.9648

See page 4 for explanation of the use of this table.

TABLE 1. THE BINOMIAL DISTRIBUTION FUNCTION

$n=8$	$r=7$	$n=9$	$r=0$	1	2	3	4	5	6	7	8
$p=0.01$		$p=0.01$	0·9135	0·9966	0·9999						
·02		·02	·8337	·9869	·9994						
·03		·03	·7602	·9718	·9980	0·9999					
·04		·04	·6925	·9522	·9955	·9997					
0·05		0·05	0·6302	0·9288	0·9916	0·9994					
·06		·06	·5730	·9022	·9862	·9987	0·9999				
·07		·07	·5204	·8729	·9791	·9977	·9998				
·08		·08	·4722	·8417	·9702	·9963	·9997				
·09		·09	·4279	·8088	·9595	·9943	·9995				
0·10		0·10	0·3874	0·7748	0·9470	0·9917	0·9991	0·9999			
·11		·11	·3504	·7401	·9328	·9883	·9986	·9999			
·12		·12	·3165	·7049	·9167	·9842	·9979	·9998			
·13		·13	·2855	·6696	·8991	·9791	·9970	·9997			
·14		·14	·2573	·6343	·8798	·9731	·9959	·9996			
0·15		0·15	0·2316	0·5995	0·8591	0·9661	0·9944	0·9994			
·16		·16	·2082	·5652	·8371	·9580	·9925	·9991	0·9999		
·17		·17	·1869	·5315	·8139	·9488	·9902	·9987	·9999		
·18		·18	·1676	·4988	·7895	·9385	·9875	·9983	·9998		
·19		·19	·1501	·4670	·7643	·9270	·9842	·9977	·9998		
0·20		0·20	0·1342	0·4362	0·7382	0·9144	0·9804	0·9969	0·9997		
·21		·21	·1199	·4066	·7115	·9006	·9760	·9960	·9996		
·22		·22	·1069	·3782	·6842	·8856	·9709	·9949	·9994		
·23		·23	·0952	·3509	·6566	·8696	·9650	·9935	·9992	0·9999	
·24		·24	·0846	·3250	·6287	·8525	·9584	·9919	·9990	·9999	
0·25		0·25	0·0751	0·3003	0·6007	0·8343	0·9511	0·9900	0·9987	0·9999	
·26		·26	·0665	·2770	·5727	·8151	·9429	·9878	·9983	·9999	
·27		·27	·0589	·2548	·5448	·7950	·9338	·9851	·9978	·9998	
·28		·28	·0520	·2340	·5171	·7740	·9238	·9821	·9972	·9997	
·29	0·9999	·29	·0458	·2144	·4898	·7522	·9130	·9787	·9965	·9997	
0·30	0·9999	0·30	0·0404	0·1960	0·4628	0·7297	0·9012	0·9747	0·9957	0·9996	
·31	·9999	·31	·0355	·1788	·4364	·7065	·8885	·9702	·9947	·9994	
·32	·9999	·32	·0311	·1628	·4106	·6827	·8748	·9652	·9936	·9993	
·33	·9999	·33	·0272	·1478	·3854	·6585	·8602	·9596	·9922	·9991	
·34	·9998	·34	·0238	·1339	·3610	·6338	·8447	·9533	·9906	·9989	0·9999
0·35	0·9998	0·35	0·0207	0·1211	0·3373	0·6089	0·8283	0·9464	0·9888	0·9986	0·9999
·36	·9997	·36	·0180	·1092	·3144	·5837	·8110	·9388	·9867	·9983	·9999
·37	·9996	·37	·0156	·0983	·2924	·5584	·7928	·9304	·9843	·9979	·9999
·38	·9996	·38	·0135	·0882	·2713	·5331	·7738	·9213	·9816	·9974	·9998
·39	·9995	·39	·0117	·0790	·2511	·5078	·7540	·9114	·9785	·9969	·9998
0·40	0·9993	0·40	0·0101	0·0705	0·2318	0·4826	0·7334	0·9006	0·9750	0·9962	0·9997
·41	·9992	·41	·0087	·0628	·2134	·4576	·7122	·8891	·9710	·9954	·9997
·42	·9990	·42	·0074	·0558	·1961	·4330	·6903	·8767	·9666	·9945	·9996
·43	·9988	·43	·0064	·0495	·1796	·4087	·6678	·8634	·9617	·9935	·9995
·44	·9986	·44	·0054	·0437	·1641	·3848	·6449	·8492	·9563	·9923	·9994
0·45	0·9983	0·45	0·0046	0·0385	0·1495	0·3614	0·6214	0·8342	0·9502	0·9909	0·9992
·46	·9980	·46	·0039	·0338	·1358	·3386	·5976	·8183	·9436	·9893	·9991
·47	·9976	·47	·0033	·0296	·1231	·3164	·5735	·8015	·9363	·9875	·9989
·48	·9972	·48	·0028	·0259	·1111	·2948	·5491	·7839	·9283	·9855	·9986
·49	·9967	·49	·0023	·0225	·1001	·2740	·5246	·7654	·9196	·9831	·9984
0·50	0·9961	0·50	0·0020	0·0195	0·0898	0·2539	0·5000	0·7461	0·9102	0·9805	0·9980

See page 4 for explanation of the use of this table.

TABLE 1. THE BINOMIAL DISTRIBUTION FUNCTION

$n = 10$	$r = 0$	1	2	3	4	5	6	7	8	9
$p = 0.01$	0.9044	0.9957	0.9999							
.02	.8171	.9838	.9991							
.03	.7374	.9655	.9972	0.9999						
.04	.6648	.9418	.9938	.9996						
0.05	0.5987	0.9139	0.9885	0.9990	0.9999					
.06	.5386	.8824	.9812	.9980	.9998					
.07	.4840	.8483	.9717	.9964	.9997					
.08	.4344	.8121	.9599	.9942	.9994					
.09	.3894	.7746	.9460	.9912	.9990	0.9999				
0.10	0.3487	0.7361	0.9298	0.9872	0.9984	0.9999				
.11	.3118	.6972	.9116	.9822	.9975	.9997				
.12	.2785	.6583	.8913	.9761	.9963	.9996				
.13	.2484	.6196	.8692	.9687	.9947	.9994	0.9999			
.14	.2213	.5816	.8455	.9600	.9927	.9990	.9999			
0.15	0.1969	0.5443	0.8202	0.9500	0.9901	0.9986	0.9999			
.16	.1749	.5080	.7936	.9386	.9870	.9980	.9998			
.17	.1552	.4730	.7659	.9259	.9832	.9973	.9997			
.18	.1374	.4392	.7372	.9117	.9787	.9963	.9996			
.19	.1216	.4068	.7078	.8961	.9734	.9951	.9994	0.9999		
0.20	0.1074	0.3758	0.6778	0.8791	0.9672	0.9936	0.9991	0.9999		
.21	.0947	.3464	.6474	.8609	.9601	.9918	.9988	.9999		
.22	.0834	.3185	.6169	.8413	.9521	.9896	.9984	.9998		
.23	.0733	.2921	.5863	.8206	.9431	.9870	.9979	.9998		
.24	.0643	.2673	.5558	.7988	.9330	.9839	.9973	.9997		
0.25	0.0563	0.2440	0.5256	0.7759	0.9219	0.9803	0.9965	0.9996		
.26	.0492	.2222	.4958	.7521	.9096	.9761	.9955	.9994		
.27	.0430	.2019	.4665	.7274	.8963	.9713	.9944	.9993	0.9999	
.28	.0374	.1830	.4378	.7021	.8819	.9658	.9930	.9990	.9999	
.29	.0326	.1655	.4099	.6761	.8663	.9596	.9913	.9988	.9999	
0.30	0.0282	0.1493	0.3828	0.6496	0.8497	0.9527	0.9894	0.9984	0.9999	
.31	.0245	.1344	.3566	.6228	.8321	.9449	.9871	.9980	.9998	
.32	.0211	.1206	.3313	.5956	.8133	.9363	.9845	.9975	.9997	
.33	.0182	.1080	.3070	.5684	.7936	.9268	.9815	.9968	.9997	
.34	.0157	.0965	.2838	.5411	.7730	.9164	.9780	.9961	.9996	
0.35	0.0135	0.0860	0.2616	0.5138	0.7515	0.9051	0.9740	0.9952	0.9995	
.36	.0115	.0764	.2405	.4868	.7292	.8928	.9695	.9941	.9993	
.37	.0098	.0677	.2206	.4600	.7061	.8795	.9644	.9929	.9991	
.38	.0084	.0598	.2017	.4336	.6823	.8652	.9587	.9914	.9989	0.9999
.39	.0071	.0527	.1840	.4077	.6580	.8500	.9523	.9897	.9986	.9999
0.40	0.0060	0.0464	0.1673	0.3823	0.6331	0.8338	0.9452	0.9877	0.9983	0.9999
.41	.0051	.0406	.1517	.3575	.6078	.8166	.9374	.9854	.9979	.9999
.42	.0043	.0355	.1372	.3335	.5822	.7984	.9288	.9828	.9975	.9998
.43	.0036	.0309	.1236	.3102	.5564	.7793	.9194	.9798	.9969	.9998
.44	.0030	.0269	.1111	.2877	.5304	.7593	.9092	.9764	.9963	.9997
0.45	0.0025	0.0233	0.0996	0.2660	0.5044	0.7384	0.8980	.9726	.9955	0.9997
.46	.0021	.0201	.0889	.2453	.4784	.7168	.8859	.9683	.9946	.9996
.47	.0017	.0173	.0791	.2255	.4526	.6943	.8729	.9634	.9935	.9995
.48	.0014	.0148	.0702	.2067	.4270	.6712	.8590	.9580	.9923	.9994
.49	.0012	.0126	.0621	.1888	.4018	.6474	.8440	.9520	.9909	.9992
0.50	0.0010	0.0107	0.0547	0.1719	0.3770	0.6230	0.8281	0.9453	0.9893	0.9990

See page 4 for explanation of the use of this table.

TABLE 1. THE BINOMIAL DISTRIBUTION FUNCTION

$n = 11$	$r = 0$	1	2	3	4	5	6	7	8	9	10
$p = 0.01$	0.8953	0.9948	0.9998								
·02	·8007	·9805	·9988								
·03	·7153	·9587	·9963	0.9998							
·04	·6382	·9308	·9917	·9993							
0·05	0·5688	0·8981	0·9848	0·9984	0·9999						
·06	·5063	·8618	·9752	·9970	·9997						
·07	·4501	·8228	·9630	·9947	·9995						
·08	·3996	·7819	·9481	·9915	·9990	0·9999					
·09	·3544	·7399	·9305	·9871	·9983	·9998					
0·10	0·3138	0·6974	0·9104	0·9815	0·9972	0·9997					
·11	·2775	·6548	·8880	·9744	·9958	·9995					
·12	·2451	·6127	·8634	·9659	·9939	·9992	0·9999				
·13	·2161	·5714	·8368	·9558	·9913	·9988	·9999				
·14	·1903	·5311	·8085	·9440	·9881	·9982	·9998				
0·15	0·1673	0·4922	0·7788	0·9306	0·9841	0·9973	0·9997				
·16	·1469	·4547	·7479	·9154	·9793	·9963	·9995				
·17	·1288	·4189	·7161	·8987	·9734	·9949	·9993	0·9999			
·18	·1127	·3849	·6836	·8803	·9666	·9932	·9990	·9999			
·19	·0985	·3526	·6506	·8603	·9587	·9910	·9986	·9998			
0·20	0·0859	0·3221	0·6174	0·8389	0·9496	0·9883	0·9980	0·9998			
·21	·0748	·2935	·5842	·8160	·9393	·9852	·9973	·9997			
·22	·0650	·2667	·5512	·7919	·9277	·9814	·9965	·9995			
·23	·0564	·2418	·5186	·7667	·9149	·9769	·9954	·9993	0·9999		
·24	·0489	·2186	·4866	·7404	·9008	·9717	·9941	·9991	·9999		
0·25	0·0422	0·1971	0·4552	0·7133	0·8854	0·9657	0·9924	0·9988	0·9999		
·26	·0364	·1773	·4247	·6854	·8687	·9588	·9905	·9984	·9998		
·27	·0314	·1590	·3951	·6570	·8507	·9510	·9881	·9979	·9998		
·28	·0270	·1423	·3665	·6281	·8315	·9423	·9854	·9973	·9997		
·29	·0231	·1270	·3390	·5989	·8112	·9326	·9821	·9966	·9996		
0·30	0·0198	0·1130	0·3127	0·5696	0·7897	0·9218	0·9784	0·9957	0·9994		
·31	·0169	·1003	·2877	·5402	·7672	·9099	·9740	·9946	·9992	0·9999	
·32	·0144	·0888	·2639	·5110	·7437	·8969	·9691	·9933	·9990	·9999	
·33	·0122	·0784	·2413	·4821	·7193	·8829	·9634	·9918	·9987	·9999	
·34	·0104	·0690	·2201	·4536	·6941	·8676	·9570	·9899	·9984	·9998	
0·35	0·0088	0·0606	0·2001	0·4256	0·6683	0·8513	0·9499	0·9878	0·9980	0·9998	
·36	·0074	·0530	·1814	·3981	·6419	·8339	·9419	·9852	·9974	·9997	
·37	·0062	·0463	·1640	·3714	·6150	·8153	·9330	·9823	·9968	·9996	
·38	·0052	·0403	·1478	·3455	·5878	·7957	·9232	·9790	·9961	·9995	
·39	·0044	·0350	·1328	·3204	·5603	·7751	·9124	·9751	·9952	·9994	
0·40	0·0036	0·0302	0·1189	0·2963	0·5328	0·7535	0·9006	0·9707	0·9941	0·9993	
·41	·0030	·0261	·1062	·2731	·5052	·7310	·8879	·9657	·9928	·9991	0·9999
·42	·0025	·0224	·0945	·2510	·4777	·7076	·8740	·9601	·9913	·9988	·9999
·43	·0021	·0192	·0838	·2300	·4505	·6834	·8592	·9539	·9896	·9986	·9999
·44	·0017	·0164	·0740	·2100	·4236	·6586	·8432	·9468	·9875	·9982	·9999
0·45	0·0014	0·0139	0·0652	0·1911	0·3971	0·6331	0·8262	0·9390	0·9852	0·9978	0·9998
·46	·0011	·0118	·0572	·1734	·3712	·6071	·8081	·9304	·9825	·9973	·9998
·47	·0009	·0100	·0501	·1567	·3459	·5807	·7890	·9209	·9794	·9967	·9998
·48	·0008	·0084	·0436	·1412	·3213	·5540	·7688	·9105	·9759	·9960	·9997
·49	·0006	·0070	·0378	·1267	·2974	·5271	·7477	·8991	·9718	·9951	·9996
0·50	0·0005	0·0059	0·0327	0·1133	0·2744	0·5000	0·7256	0·8867	0·9673	0·9941	0·9995

See page 4 for explanation of the use of this table.

TABLE 1. THE BINOMIAL DISTRIBUTION FUNCTION

$n = 12$	$r = 0$	1	2	3	4	5	6	7	8	9	10
$p = 0.01$	0.8864	0.9938	0.9998								
.02	.7847	.9769	.9985	0.9999							
.03	.6938	.9514	.9952	.9997							
.04	.6127	.9191	.9893	.9990	0.9999						
0.05	0.5404	0.8816	0.9804	0.9978	0.9998						
.06	.4759	.8405	.9684	.9957	.9996						
.07	.4186	.7967	.9532	.9925	.9991	0.9999					
.08	.3677	.7513	.9348	.9880	.9984	.9998					
.09	.3225	.7052	.9134	.9820	.9973	.9997					
0.10	0.2824	0.6590	0.8891	0.9744	0.9957	0.9995	0.9999				
.11	.2470	.6133	.8623	.9649	.9935	.9991	.9999				
.12	.2157	.5686	.8333	.9536	.9905	.9986	.9998				
.13	.1880	.5252	.8023	.9403	.9867	.9978	.9997				
.14	.1637	.4834	.7697	.9250	.9819	.9967	.9996				
0.15	0.1422	0.4435	0.7358	0.9078	0.9761	0.9954	0.9993	0.9999			
.16	.1234	.4055	.7010	.8886	.9690	.9935	.9990	.9999			
.17	.1069	.3696	.6656	.8676	.9607	.9912	.9985	.9998			
.18	.0924	.3359	.6298	.8448	.9511	.9884	.9979	.9997			
.19	.0798	.3043	.5940	.8205	.9400	.9849	.9971	.9996			
0.20	0.0687	0.2749	0.5583	0.7946	0.9274	0.9806	0.9961	0.9994	0.9999		
.21	.0591	.2476	.5232	.7674	.9134	.9755	.9948	.9992	.9999		
.22	.0507	.2224	.4886	.7390	.8979	.9696	.9932	.9989	.9999		
.23	.0434	.1991	.4550	.7096	.8808	.9626	.9911	.9984	.9998		
.24	.0371	.1778	.4222	.6795	.8623	.9547	.9887	.9979	.9997		
0.25	0.0317	0.1584	0.3907	0.6488	0.8424	0.9456	0.9857	0.9972	0.9996		
.26	.0270	.1406	.3603	.6176	.8210	.9354	.9822	.9964	.9995	0.9999	
.27	.0229	.1245	.3313	.5863	.7984	.9240	.9781	.9953	.9993	.9999	
.28	.0194	.1100	.3037	.5548	.7746	.9113	.9733	.9940	.9990	.9999	
.29	.0164	.0968	.2775	.5235	.7496	.8974	.9678	.9924	.9987	.9998	
0.30	0.0138	0.0850	0.2528	0.4925	0.7237	0.8822	0.9614	0.9905	0.9983	0.9998	
.31	.0116	.0744	.2296	.4619	.6968	.8657	.9542	.9882	.9978	.9997	
.32	.0098	.0650	.2078	.4319	.6692	.8479	.9460	.9856	.9972	.9996	
.33	.0082	.0565	.1876	.4027	.6410	.8289	.9368	.9824	.9964	.9995	
.34	.0068	.0491	.1687	.3742	.6124	.8087	.9266	.9787	.9955	.9993	0.9999
0.35	0.0057	0.0424	0.1513	0.3467	0.5833	0.7873	0.9154	0.9745	0.9944	0.9992	0.9999
.36	.0047	.0366	.1352	.3201	.5541	.7648	.9030	.9696	.9930	.9989	.9999
.37	.0039	.0315	.1205	.2947	.5249	.7412	.8894	.9641	.9915	.9986	.9999
.38	.0032	.0270	.1069	.2704	.4957	.7167	.8747	.9578	.9896	.9982	.9998
.39	.0027	.0230	.0946	.2472	.4668	.6913	.8589	.9507	.9873	.9978	.9998
0.40	0.0022	0.0196	0.0834	0.2253	0.4382	0.6652	0.8418	0.9427	0.9847	0.9972	0.9997
.41	.0018	.0166	.0733	.2047	.4101	.6384	.8235	.9338	.9817	.9965	.9996
.42	.0014	.0140	.0642	.1853	.3825	.6111	.8041	.9240	.9782	.9957	.9995
.43	.0012	.0118	.0560	.1671	.3557	.5833	.7836	.9131	.9742	.9947	.9993
.44	.0010	.0099	.0487	.1502	.3296	.5552	.7620	.9012	.9696	.9935	.9991
0.45	0.0008	0.0083	0.0421	0.1345	0.3044	0.5269	0.7393	0.8883	0.9644	0.9921	0.9989
.46	.0006	.0069	.0363	.1199	.2802	.4986	.7157	.8742	.9585	.9905	.9986
.47	.0005	.0057	.0312	.1066	.2570	.4703	.6911	.8589	.9519	.9886	.9983
.48	.0004	.0047	.0267	.0943	.2348	.4423	.6657	.8425	.9445	.9863	.9979
.49	.0003	.0039	.0227	.0832	.2138	.4145	.6396	.8249	.9362	.9837	.9974
0.50	0.0002	0.0032	0.0193	0.0730	0.1938	0.3872	0.6128	0.8062	0.9270	0.9807	0.9968

$n = 12$	$r = 11$
$p = 0.44$	0.9999
0.45	0.9999
.46	.9999
.47	.9999
.48	.9999
.49	.9998
0.50	0.9998

See page 4 for explanation of the use of this table.

TABLE 1. THE BINOMIAL DISTRIBUTION FUNCTION

$n = 13$	$r = 0$	1	2	3	4	5	6	7	8	9	10
$p = 0.01$	0.8775	0.9928	0.9997								
.02	.7690	.9730	.9980	.9999							
.03	.6730	.9436	.9938	.9995							
.04	.5882	.9068	.9865	.9986	0.9999						
0.05	0.5133	0.8646	0.9755	0.9969	0.9997						
.06	.4474	.8186	.9608	.9940	.9993	0.9999					
.07	.3893	.7702	.9422	.9897	.9987	.9999					
.08	.3383	.7206	.9201	.9837	.9976	.9997					
.09	.2935	.6707	.8946	.9758	.9959	.9995	0.9999				
0.10	0.2542	0.6213	0.8661	0.9658	0.9935	0.9991	0.9999				
.11	.2198	.5730	.8349	.9536	.9903	.9985	.9998				
.12	.1898	.5262	.8015	.9391	.9861	.9976	.9997				
.13	.1636	.4814	.7663	.9224	.9807	.9964	.9995	0.9999			
.14	.1408	.4386	.7296	.9033	.9740	.9947	.9992	.9999			
0.15	0.1209	0.3983	0.6920	0.8820	0.9658	0.9925	0.9987	0.9998			
.16	.1037	.3604	.6537	.8586	.9562	.9896	.9981	.9997			
.17	.0887	.3249	.6152	.8333	.9449	.9861	.9973	.9996			
.18	.0758	.2920	.5769	.8061	.9319	.9817	.9962	.9994	0.9999		
.19	.0646	.2616	.5389	.7774	.9173	.9763	.9948	.9991	.9999		
0.20	0.0550	0.2336	0.5017	0.7473	0.9009	0.9700	0.9930	0.9988	0.9998		
.21	.0467	.2080	.4653	.7161	.8827	.9625	.9907	.9983	.9998		
.22	.0396	.1846	.4301	.6839	.8629	.9538	.9880	.9976	.9996		
.23	.0334	.1633	.3961	.6511	.8415	.9438	.9846	.9968	.9995	0.9999	
.24	.0282	.1441	.3636	.6178	.8184	.9325	.9805	.9957	.9993	.9999	
0.25	0.0238	0.1267	0.3326	0.5843	0.7940	0.9198	0.9757	0.9944	0.9990	0.9999	
.26	.0200	.1111	.3032	.5507	.7681	.9056	.9701	.9927	.9987	.9998	
.27	.0167	.0971	.2755	.5174	.7411	.8901	.9635	.9907	.9982	.9997	
.28	.0140	.0846	.2495	.4845	.7130	.8730	.9560	.9882	.9976	.9996	
.29	.0117	.0735	.2251	.4522	.6840	.8545	.9473	.9853	.9969	.9995	0.9999
0.30	0.0097	0.0637	0.2025	0.4206	0.6543	0.8346	0.9376	0.9818	0.9960	0.9993	0.9999
.31	.0080	.0550	.1815	.3899	.6240	.8133	.9267	.9777	.9948	.9991	.9999
.32	.0066	.0473	.1621	.3602	.5933	.7907	.9146	.9729	.9935	.9988	.9999
.33	.0055	.0406	.1443	.3317	.5624	.7669	.9012	.9674	.9918	.9985	.9998
.34	.0045	.0347	.1280	.3043	.5314	.7419	.8865	.9610	.9898	.9980	.9997
0.35	0.0037	0.0296	0.1132	0.2783	0.5005	0.7159	0.8705	0.9538	0.9874	0.9975	0.9997
.36	.0030	.0251	.0997	.2536	.4699	.6889	.8532	.9456	.9846	.9968	.9995
.37	.0025	.0213	.0875	.2302	.4397	.6612	.8346	.9365	.9813	.9960	.9994
.38	.0020	.0179	.0765	.2083	.4101	.6327	.8147	.9262	.9775	.9949	.9992
.39	.0016	.0151	.0667	.1877	.3812	.6038	.7935	.9149	.9730	.9937	.9990
0.40	0.0013	0.0126	0.0579	0.1686	0.3530	0.5744	0.7712	0.9023	0.9679	0.9922	0.9987
.41	.0010	.0105	.0501	.1508	.3258	.5448	.7476	.8886	.9621	.9904	.9983
.42	.0008	.0088	.0431	.1344	.2997	.5151	.7230	.8736	.9554	.9883	.9979
.43	.0007	.0072	.0370	.1193	.2746	.4854	.6975	.8574	.9480	.9859	.9973
.44	.0005	.0060	.0316	.1055	.2507	.4559	.6710	.8400	.9395	.9830	.9967
0.45	0.0004	0.0049	0.0269	0.0929	0.2279	0.4268	0.6437	0.8212	0.9302	0.9797	0.9959
.46	.0003	.0040	.0228	.0815	.2065	.3981	.6158	.8012	.9197	.9758	.9949
.47	.0003	.0033	.0192	.0712	.1863	.3701	.5873	.7800	.9082	.9713	.9937
.48	.0002	.0026	.0162	.0619	.1674	.3427	.5585	.7576	.8955	.9662	.9923
.49	.0002	.0021	.0135	.0536	.1498	.3162	.5293	.7341	.8817	.9604	.9907
0.50	0.0001	0.0017	0.0112	0.0461	0.1334	0.2905	0.5000	0.7095	0.8666	0.9539	0.9888

See page 4 for explanation of the use of this table.

TABLE 1. THE BINOMIAL DISTRIBUTION FUNCTION

$n=13$	$r=11$	12		$n=14$	$r=0$	1	2	3	4	5	6	7
$p=0.01$				$p=0.01$	0·8687	0·9916	0·9997					
·02				·02	·7536	·9690	·9975	0·9999				
·03				·03	·6528	·9355	·9923	·9994				
·04				·04	·5647	·8941	·9833	·9981	0·9998			
0·05				0·05	0·4877	0·8470	0·9699	0·9958	0·9996			
·06				·06	·4205	·7963	·9522	·9920	·9990	0·9999		
·07				·07	·3620	·7436	·9302	·9864	·9980	·9998		
·08				·08	·3112	·6900	·9042	·9786	·9965	·9996		
·09				·09	·2670	·6368	·8745	·9685	·9941	·9992	0·9999	
0·10				0·10	0·2288	0·5846	0·8416	0·9559	0·9908	0·9985	0·9998	
·11				·11	·1956	·5342	·8061	·9406	·9863	·9976	·9997	
·12				·12	·1670	·4859	·7685	·9226	·9804	·9962	·9994	
·13				·13	·1423	·4401	·7292	·9021	·9731	·9943	·9991	0·9999
·14				·14	·1211	·3969	·6889	·8790	·9641	·9918	·9985	·9998
0·15				0·15	0·1028	0·3567	0·6479	0·8535	0·9533	0·9885	0·9978	0·9997
·16				·16	·0871	·3193	·6068	·8258	·9406	·9843	·9968	·9995
·17				·17	·0736	·2848	·5659	·7962	·9259	·9791	·9954	·9992
·18				·18	·0621	·2531	·5256	·7649	·9093	·9727	·9936	·9988
·19				·19	·0523	·2242	·4862	·7321	·8907	·9651	·9913	·9983
0·20				0·20	0·0440	0·1979	0·4481	0·6982	0·8702	0·9561	0·9884	0·9976
·21				·21	·0369	·1741	·4113	·6634	·8477	·9457	·9848	·9967
·22				·22	·0309	·1527	·3761	·6281	·8235	·9338	·9804	·9955
·23				·23	·0258	·1335	·3426	·5924	·7977	·9203	·9752	·9940
·24				·24	·0214	·1163	·3109	·5568	·7703	·9051	·9690	·9921
0·25				0·25	0·0178	0·1010	0·2811	0·5213	0·7415	0·8883	0·9617	0·9897
·26				·26	·0148	·0874	·2533	·4864	·7116	·8699	·9533	·9868
·27				·27	·0122	·0754	·2273	·4521	·6807	·8498	·9437	·9833
·28				·28	·0101	·0648	·2033	·4187	·6491	·8282	·9327	·9792
·29				·29	·0083	·0556	·1812	·3863	·6168	·8051	·9204	·9743
0·30				0·30	0·0068	0·0475	0·1608	0·3552	0·5842	0·7805	0·9067	0·9685
·31				·31	·0055	·0404	·1423	·3253	·5514	·7546	·8916	·9619
·32				·32	·0045	·0343	·1254	·2968	·5187	·7276	·8750	·9542
·33				·33	·0037	·0290	·1101	·2699	·4862	·6994	·8569	·9455
·34				·34	·0030	·0244	·0963	·2444	·4542	·6703	·8374	·9357
0·35				0·35	0·0024	0·0205	0·0839	0·2205	0·4227	0·6405	0·8164	0·9247
·36				·36	·0019	·0172	·0729	·1982	·3920	·6101	·7941	·9124
·37	0·9999			·37	·0016	·0143	·0630	·1774	·3622	·5792	·7704	·8988
·38	·9999			·38	·0012	·0119	·0543	·1582	·3334	·5481	·7455	·8838
·39	·9999			·39	·0010	·0098	·0466	·1405	·3057	·5169	·7195	·8675
0·40	0·9999			0·40	0·0008	0·0081	0·0398	0·1243	0·2793	0·4859	0·6925	0·8499
·41	·9998			·41	·0006	·0066	·0339	·1095	·2541	·4550	·6645	·8308
·42	·9998			·42	·0005	·0054	·0287	·0961	·2303	·4246	·6357	·8104
·43	·9997			·43	·0004	·0044	·0242	·0839	·2078	·3948	·6063	·7887
·44	·9996			·44	·0003	·0036	·0203	·0730	·1868	·3656	·5764	·7656
0·45	0·9995			0·45	0·0002	0·0029	0·0170	0·0632	0·1672	0·3373	0·5461	0·7414
·46	·9993			·46	·0002	·0023	·0142	·0545	·1490	·3100	·5157	·7160
·47	·9991	0·9999		·47	·0001	·0019	·0117	·0468	·1322	·2837	·4852	·6895
·48	·9989	·9999		·48	·0001	·0015	·0097	·0399	·1167	·2585	·4549	·6620
·49	·9986	·9999		·49	·0001	·0012	·0079	·0339	·1026	·2346	·4249	·6337
0·50	0·9983	0·9999		0·50	0·0001	0·0009	0·0065	0·0287	0·0898	0·2120	0·3953	0·6047

See page 4 for explanation of the use of this table.

TABLE 1. THE BINOMIAL DISTRIBUTION FUNCTION

n = 14

p	r = 8	9	10	11	12	13
0·01						
·02						
·03						
·04						
0·05						
·06						
·07						
·08						
·09						
0·10						
·11						
·12						
·13						
·14						
0·15						
·16	·9999					
·17	·9999					
·18	·9998					
·19	·9997					
0·20	0·9996					
·21	·9994	0·9999				
·22	·9992	·9999				
·23	·9989	·9998				
·24	·9984	·9998				
0·25	0·9978	0·9997				
·26	·9971	·9995	0·9999			
·27	·9962	·9993	·9999			
·28	·9950	·9991	·9999			
·29	·9935	·9988	·9998			
0·30	0·9917	0·9983	0·9998			
·31	·9895	·9978	·9997			
·32	·9869	·9971	·9995	0·9999		
·33	·9837	·9963	·9994	·9999		
·34	·9800	·9952	·9992	·9999		
0·35	0·9757	0·9940	0·9989	0·9999		
·36	·9706	·9924	·9986	·9998		
·37	·9647	·9905	·9981	·9997		
·38	·9580	·9883	·9976	·9997		
·39	·9503	·9856	·9969	·9995		
0·40	0·9417	0·9825	0·9961	0·9994	0·9999	
·41	·9320	·9788	·9951	·9992	·9999	
·42	·9211	·9745	·9939	·9990	·9999	
·43	·9090	·9696	·9924	·9987	·9999	
·44	·8957	·9639	·9907	·9983	·9998	
0·45	0·8811	0·9574	0·9886	0·9978	0·9997	
·46	·8652	·9500	·9861	·9973	·9997	
·47	·8480	·9417	·9832	·9966	·9996	
·48	·8293	·9323	·9798	·9958	·9994	
·49	·8094	·9218	·9759	·9947	·9993	
0·50	0·7880	0·9102	0·9713	0·9935	0·9991	0·9999

n = 15

p	r = 0	1	2	3
0·01	0·8601	0·9904	0·9996	
·02	·7386	·9647	·9970	0·9998
·03	·6333	·9270	·9906	·9992
·04	·5421	·8809	·9797	·9976
0·05	0·4633	0·8290	0·9638	0·9945
·06	·3953	·7738	·9429	·9896
·07	·3367	·7168	·9171	·9825
·08	·2863	·6597	·8870	·9727
·09	·2430	·6035	·8531	·9601
0·10	0·2059	0·5490	0·8159	0·9444
·11	·1741	·4969	·7762	·9258
·12	·1470	·4476	·7346	·9041
·13	·1238	·4013	·6916	·8796
·14	·1041	·3583	·6480	·8524
0·15	0·0874	0·3186	0·6042	0·8227
·16	·0731	·2821	·5608	·7908
·17	·0611	·2489	·5181	·7571
·18	·0510	·2187	·4766	·7218
·19	·0424	·1915	·4365	·6854
0·20	0·0352	0·1671	0·3980	0·6482
·21	·0291	·1453	·3615	·6105
·22	·0241	·1259	·3269	·5726
·23	·0198	·1087	·2945	·5350
·24	·0163	·0935	·2642	·4978
0·25	0·0134	0·0802	0·2361	0·4613
·26	·0109	·0685	·2101	·4258
·27	·0089	·0583	·1863	·3914
·28	·0072	·0495	·1645	·3584
·29	·0059	·0419	·1447	·3268
0·30	0·0047	0·0353	0·1268	0·2969
·31	·0038	·0296	·1107	·2686
·32	·0031	·0248	·0962	·2420
·33	·0025	·0206	·0833	·2171
·34	·0020	·0171	·0719	·1940
0·35	0·0016	0·0142	0·0617	0·1727
·36	·0012	·0117	·0528	·1531
·37	·0010	·0096	·0450	·1351
·38	·0008	·0078	·0382	·1187
·39	·0006	·0064	·0322	·1039
0·40	0·0005	0·0052	0·0271	0·0905
·41	·0004	·0042	·0227	·0785
·42	·0003	·0034	·0189	·0678
·43	·0002	·0027	·0157	·0583
·44	·0002	·0021	·0130	·0498
0·45	0·0001	0·0017	0·0107	0·0424
·46	·0001	·0013	·0087	·0359
·47	·0001	·0010	·0071	·0303
·48	·0001	·0008	·0057	·0254
·49		·0006	·0046	·0212
0·50		0·0005	0·0037	0·0176

See page 4 for explanation of the use of this table.

TABLE 1. THE BINOMIAL DISTRIBUTION FUNCTION

$n = 15$	$r = 4$	5	6	7	8	9	10	11	12	13
$p = 0.01$										
.02										
.03	0·9999									
.04	·9998									
0·05	0·9994	0·9999								
·06	·9986	·9999								
·07	·9972	·9997								
·08	·9950	·9993	0·9999							
·09	·9918	·9987	·9998							
0·10	0·9873	0·9978	0·9997							
·11	·9813	·9963	·9994	0·9999						
·12	·9735	·9943	·9990	·9999						
·13	·9639	·9916	·9985	·9998						
·14	·9522	·9879	·9976	·9996						
0·15	0·9383	0·9832	0·9964	0·9994	0·9999					
·16	·9222	·9773	·9948	·9990	·9999					
·17	·9039	·9700	·9926	·9986	·9998					
·18	·8833	·9613	·9898	·9979	·9997					
·19	·8606	·9510	·9863	·9970	·9995	0·9999				
0·20	0·8358	0·9389	0·9819	0·9958	0·9992	0·9999				
·21	·8090	·9252	·9766	·9942	·9989	·9998				
·22	·7805	·9095	·9702	·9922	·9984	·9997				
·23	·7505	·8921	·9626	·9896	·9977	·9996	0·9999			
·24	·7190	·8728	·9537	·9865	·9969	·9994	·9999			
0·25	0·6865	0·8516	0·9434	0·9827	0·9958	0·9992	0·9999			
·26	·6531	·8287	·9316	·9781	·9944	·9989	·9998			
·27	·6190	·8042	·9183	·9726	·9927	·9985	·9998			
·28	·5846	·7780	·9035	·9662	·9906	·9979	·9997			
·29	·5500	·7505	·8870	·9587	·9879	·9972	·9995	0·9999		
0·30	0·5155	0·7216	0·8689	0·9500	0·9848	0·9963	0·9993	0·9999		
·31	·4813	·6916	·8491	·9401	·9810	·9952	·9991	·9999		
·32	·4477	·6607	·8278	·9289	·9764	·9938	·9988	·9998		
·33	·4148	·6291	·8049	·9163	·9711	·9921	·9984	·9997		
·34	·3829	·5968	·7806	·9023	·9649	·9901	·9978	·9996		
0·35	0·3519	0·5643	0·7548	0·8868	0·9578	0·9876	0·9972	0·9995	0·9999	
·36	·3222	·5316	·7278	·8698	·9496	·9846	·9963	·9994	·9999	
·37	·2938	·4989	·6997	·8513	·9403	·9810	·9953	·9991	·9999	
·38	·2668	·4665	·6705	·8313	·9298	·9768	·9941	·9989	·9998	
·39	·2413	·4346	·6405	·8098	·9180	·9719	·9925	·9985	·9998	
0·40	0·2173	0·4032	0·6098	0·7869	0·9050	0·9662	0·9907	0·9981	0·9997	
·41	·1948	·3726	·5786	·7626	·8905	·9596	·9884	·9975	·9996	
·42	·1739	·3430	·5470	·7370	·8746	·9521	·9857	·9968	·9995	
·43	·1546	·3144	·5153	·7102	·8573	·9435	·9826	·9960	·9993	0·9999
·44	·1367	·2869	·4836	·6824	·8385	·9339	·9789	·9949	·9991	·9999
0·45	0·1204	0·2608	0·4522	0·6535	0·8182	0·9231	0·9745	0·9937	0·9989	0·9999
·46	·1055	·2359	·4211	·6238	·7966	·9110	·9695	·9921	·9986	·9998
·47	·0920	·2125	·3905	·5935	·7735	·8976	·9637	·9903	·9982	·9998
·48	·0799	·1905	·3606	·5626	·7490	·8829	·9570	·9881	·9977	·9997
·49	·0690	·1699	·3316	·5314	·7233	·8667	·9494	·9855	·9971	·9996
0·50	0·0592	0·1509	0·3036	0·5000	0·6964	0·8491	0·9408	0·9824	0·9963	0·9995

See page 4 for explanation of the use of this table.

TABLE 1. THE BINOMIAL DISTRIBUTION FUNCTION

$n = 16$	$r = 0$	1	2	3	4	5	6	7	8	9	10
$p = 0.01$	0.8515	0.9891	0.9995								
.02	.7238	.9601	.9963	0.9998							
.03	.6143	.9182	.9887	.9989	0.9999						
.04	.5204	.8673	.9758	.9968	.9997						
0.05	0.4401	0.8108	0.9571	0.9930	0.9991	0.9999					
.06	.3716	.7511	.9327	.9868	.9981	.9998					
.07	.3131	.6902	.9031	.9779	.9962	.9995	0.9999				
.08	.2634	.6299	.8689	.9658	.9932	.9990	.9999				
.09	.2211	.5711	.8306	.9504	.9889	.9981	.9997				
0.10	0.1853	0.5147	0.7892	0.9316	0.9830	0.9967	0.9995	0.9999			
.11	.1550	.4614	.7455	.9093	.9752	.9947	.9991	.9999			
.12	.1293	.4115	.7001	.8838	.9652	.9918	.9985	.9998			
.13	.1077	.3653	.6539	.8552	.9529	.9880	.9976	.9996	0.9999		
.14	.0895	.3227	.6074	.8237	.9382	.9829	.9962	.9993	.9999		
0.15	0.0743	0.2839	0.5614	0.7899	0.9209	0.9765	0.9944	0.9989	0.9998		
.16	.0614	.2487	.5162	.7540	.9012	.9685	.9920	.9984	.9997		
.17	.0507	.2170	.4723	.7164	.8789	.9588	.9888	.9976	.9996	0.9999	
.18	.0418	.1885	.4302	.6777	.8542	.9473	.9847	.9964	.9993	.9999	
.19	.0343	.1632	.3899	.6381	.8273	.9338	.9796	.9949	.9990	.9998	
0.20	0.0281	0.1407	0.3518	0.5981	0.7982	0.9183	0.9733	0.9930	0.9985	0.9998	
.21	.0230	.1209	.3161	.5582	.7673	.9008	.9658	.9905	.9979	.9996	0.9999
.22	.0188	.1035	.2827	.5186	.7348	.8812	.9568	.9873	.9970	.9994	.9999
.23	.0153	.0883	.2517	.4797	.7009	.8595	.9464	.9834	.9959	.9992	.9999
.24	.0124	.0750	.2232	.4417	.6659	.8359	.9343	.9786	.9944	.9988	.9998
0.25	0.0100	0.0635	0.1971	0.4050	0.6302	0.8103	0.9204	0.9729	0.9925	0.9984	0.9997
.26	.0081	.0535	.1733	.3697	.5940	.7831	.9049	.9660	.9902	.9977	.9996
.27	.0065	.0450	.1518	.3360	.5575	.7542	.8875	.9580	.9873	.9969	.9994
.28	.0052	.0377	.1323	.3041	.5212	.7239	.8683	.9486	.9837	.9959	.9992
.29	.0042	.0314	.1149	.2740	.4853	.6923	.8474	.9379	.9794	.9945	.9989
0.30	0.0033	0.0261	0.0994	0.2459	0.4499	0.6598	0.8247	0.9256	0.9743	0.9929	0.9984
.31	.0026	.0216	.0856	.2196	.4154	.6264	.8003	.9119	.9683	.9908	.9979
.32	.0021	.0178	.0734	.1953	.3819	.5926	.7743	.8965	.9612	.9883	.9972
.33	.0016	.0146	.0626	.1730	.3496	.5584	.7469	.8795	.9530	.9852	.9963
.34	.0013	.0120	.0533	.1525	.3187	.5241	.7181	.8609	.9436	.9815	.9952
0.35	0.0010	0.0098	0.0451	0.1339	0.2892	0.4900	0.6881	0.8406	0.9329	0.9771	0.9938
.36	.0008	.0079	.0380	.1170	.2613	.4562	.6572	.8187	.9209	.9720	.9921
.37	.0006	.0064	.0319	.1018	.2351	.4230	.6254	.7952	.9074	.9659	.9900
.38	.0005	.0052	.0266	.0881	.2105	.3906	.5930	.7702	.8924	.9589	.9875
.39	.0004	.0041	.0222	.0759	.1877	.3592	.5602	.7438	.8758	.9509	.9845
0.40	0.0003	0.0033	0.0183	0.0651	0.1666	0.3288	0.5272	0.7161	0.8577	0.9417	0.9809
.41	.0002	.0026	.0151	.0556	.1471	.2997	.4942	.6872	.8381	.9313	.9766
.42	.0002	.0021	.0124	.0473	.1293	.2720	.4613	.6572	.8168	.9195	.9716
.43	.0001	.0016	.0101	.0400	.1131	.2457	.4289	.6264	.7940	.9064	.9658
.44	.0001	.0013	.0082	.0336	.0985	.2208	.3971	.5949	.7698	.8919	.9591
0.45	0.0001	0.0010	0.0066	0.0281	0.0853	0.1976	0.3660	0.5629	0.7441	0.8759	0.9514
.46	.0001	.0008	.0053	.0234	.0735	.1759	.3359	.5306	.7171	.8584	.9426
.47		.0006	.0042	.0194	.0630	.1559	.3068	.4981	.6889	.8393	.9326
.48		.0005	.0034	.0160	.0537	.1374	.2790	.4657	.6595	.8186	.9214
.49		.0003	.0027	.0131	.0456	.1205	.2524	.4335	.6293	.7964	.9089
0.50		0.0003	0.0021	0.0106	0.0384	0.1051	0.2272	0.4018	0.5982	0.7728	0.8949

See page 4 for explanation of the use of this table.

TABLE 1. THE BINOMIAL DISTRIBUTION FUNCTION

$n=16$ p	$r=11$	12	13	14	$n=17$ p	$r=0$	1	2	3	4	5
0·01					0·01	0·8429	0·9877	0·9994			
·02					·02	·7093	·9554	·9956	·9997		
·03					·03	·5958	·9091	·9866	·9986	0·9999	
·04					·04	·4996	·8535	·9714	·9960	·9996	
0·05					0·05	0·4181	·7922	·9497	·9912	·9988	0·9999
·06					·06	·3493	·7283	·9218	·9836	·9974	·9997
·07					·07	·2912	·6638	·8882	·9727	·9949	·9993
·08					·08	·2423	·6005	·8497	·9581	·9911	·9985
·09					·09	·2012	·5396	·8073	·9397	·9855	·9973
0·10					0·10	0·1668	0·4818	0·7618	0·9174	0·9779	0·9953
·11					·11	·1379	·4277	·7142	·8913	·9679	·9925
·12					·12	·1138	·3777	·6655	·8617	·9554	·9886
·13					·13	·0937	·3318	·6164	·8290	·9402	·9834
·14					·14	·0770	·2901	·5676	·7935	·9222	·9766
0·15					0·15	0·0631	0·2525	0·5198	0·7556	0·9013	0·9681
·16					·16	·0516	·2187	·4734	·7159	·8776	·9577
·17					·17	·0421	·1887	·4289	·6749	·8513	·9452
·18					·18	·0343	·1621	·3867	·6331	·8225	·9305
·19					·19	·0278	·1387	·3468	·5909	·7913	·9136
0·20					0·20	0·0225	0·1182	0·3096	0·5489	0·7582	0·8943
·21					·21	·0182	·1004	·2751	·5073	·7234	·8727
·22					·22	·0146	·0849	·2433	·4667	·6872	·8490
·23					·23	·0118	·0715	·2141	·4272	·6500	·8230
·24					·24	·0094	·0600	·1877	·3893	·6121	·7951
0·25					0·25	0·0075	0·0501	0·1637	0·3530	0·5739	0·7653
·26	0·9999				·26	·0060	·0417	·1422	·3186	·5357	·7339
·27	·9999				·27	·0047	·0346	·1229	·2863	·4977	·7011
·28	·9999				·28	·0038	·0286	·1058	·2560	·4604	·6671
·29	·9998				·29	·0030	·0235	·0907	·2279	·4240	·6323
0·30	0·9997				0·30	0·0023	0·0193	0·0774	0·2019	0·3887	0·5968
·31	·9996				·31	·0018	·0157	·0657	·1781	·3547	·5610
·32	·9995	0·9999			·32	·0014	·0128	·0556	·1563	·3222	·5251
·33	·9993	·9999			·33	·0011	·0104	·0468	·1366	·2913	·4895
·34	·9990	·9999			·34	·0009	·0083	·0392	·1188	·2622	·4542
0·35	0·9987	·9998			0·35	0·0007	0·0067	0·0327	0·1028	0·2348	0·4197
·36	·9983	·9997			·36	·0005	·0054	·0272	·0885	·2094	·3861
·37	·9977	·9996			·37	·0004	·0043	·0225	·0759	·1858	·3535
·38	·9970	·9995	0·9999		·38	·0003	·0034	·0185	·0648	·1640	·3222
·39	·9962	·9993	·9999		·39	·0002	·0027	·0151	·0550	·1441	·2923
0·40	0·9951	·9991	·9999		0·40	0·0002	0·0021	0·0123	0·0464	0·1260	0·2639
·41	·9938	·9988	·9998		·41	·0001	·0016	·0100	·0390	·1096	·2372
·42	·9922	·9984	·9998		·42	·0001	·0013	·0080	·0326	·0949	·2121
·43	·9902	·9979	·9997		·43	·0001	·0010	·0065	·0271	·0817	·1887
·44	·9879	·9973	·9996		·44	·0001	·0008	·0052	·0224	·0699	·1670
0·45	0·9851	·9965	·9994	0·9999	0·45		0·0006	0·0041	0·0184	0·0596	0·1471
·46	·9817	·9956	·9993	·9999	·46		·0004	·0032	·0151	·0505	·1288
·47	·9778	·9945	·9990	·9999	·47		·0003	·0025	·0123	·0425	·1122
·48	·9732	·9931	·9987	·9999	·48		·0002	·0020	·0099	·0356	·0972
·49	·9678	·9914	·9984	·9998	·49		·0002	·0015	·0080	·0296	·0838
0·50	0·9616	0·9894	0·9979	0·9997	0·50		0·0001	0·0012	0·0064	0·0245	0·0717

See page 4 for explanation of the use of this table.

TABLE 1. THE BINOMIAL DISTRIBUTION FUNCTION

$n = 17$	$r = 6$	7	8	9	10	11	12	13	14	15
$p = 0.01$										
.02										
.03										
.04										
0.05										
.06										
.07	0.9999									
.08	.9998									
.09	.9996									
0.10	0.9992	0.9999								
.11	.9986	.9998								
.12	.9977	.9996	0.9999							
.13	.9963	.9993	.9999							
.14	.9944	.9989	.9998							
0.15	0.9917	0.9983	0.9997							
.16	.9882	.9973	.9995	0.9999						
.17	.9837	.9961	.9992	.9999						
.18	.9780	.9943	.9988	.9998						
.19	.9709	.9920	.9982	.9997						
0.20	0.9623	0.9891	0.9974	0.9995	0.9999					
.21	.9521	.9853	.9963	.9993	.9999					
.22	.9402	.9806	.9949	.9989	.9998					
.23	.9264	.9749	.9930	.9984	.9997					
.24	.9106	.9680	.9906	.9978	.9996	0.9999				
0.25	0.8929	0.9598	0.9876	0.9969	0.9994	0.9999				
.26	.8732	.9501	.9839	.9958	.9991	.9998				
.27	.8515	.9389	.9794	.9943	.9987	.9998				
.28	.8279	.9261	.9739	.9925	.9982	.9997				
.29	.8024	.9116	.9674	.9902	.9976	.9995	0.9999			
0.30	0.7752	0.8954	0.9597	0.9873	0.9968	0.9993	0.9999			
.31	.7464	.8773	.9508	.9838	.9957	.9991	.9998			
.32	.7162	.8574	.9405	.9796	.9943	.9987	.9998			
.33	.6847	.8358	.9288	.9746	.9926	.9983	.9997			
.34	.6521	.8123	.9155	.9686	.9905	.9977	.9996	0.9999		
0.35	0.6188	0.7872	0.9006	0.9617	0.9880	0.9970	0.9994	0.9999		
.36	.5848	.7605	.8841	.9536	.9849	.9960	.9992	.9999		
.37	.5505	.7324	.8659	.9443	.9811	.9949	.9989	.9998		
.38	.5161	.7029	.8459	.9336	.9766	.9934	.9985	.9998		
.39	.4818	.6722	.8243	.9216	.9714	.9916	.9981	.9997		
0.40	0.4478	0.6405	0.8011	0.9081	0.9652	0.9894	0.9975	0.9995	0.9999	
.41	.4144	.6080	.7762	.8930	.9580	.9867	.9967	.9994	.9999	
.42	.3818	.5750	.7498	.8764	.9497	.9835	.9958	.9992	.9999	
.43	.3501	.5415	.7220	.8581	.9403	.9797	.9946	.9989	.9998	
.44	.3195	.5079	.6928	.8382	.9295	.9752	.9931	.9986	.9998	
0.45	0.2902	0.4743	0.6626	0.8166	0.9174	0.9699	0.9914	0.9981	0.9997	
.46	.2623	.4410	.6313	.7934	.9038	.9637	.9892	.9976	.9996	
.47	.2359	.4082	.5992	.7686	.8888	.9566	.9866	.9969	.9995	0.9999
.48	.2110	.3761	.5665	.7423	.8721	.9483	.9835	.9960	.9993	.9999
.49	.1878	.3448	.5333	.7145	.8538	.9389	.9798	.9950	.9991	.9999
0.50	0.1662	0.3145	0.5000	0.6855	0.8338	0.9283	0.9755	0.9936	0.9988	0.9999

See page 4 for explanation of the use of this table.

TABLE 1. THE BINOMIAL DISTRIBUTION FUNCTION

$n = 18$	$r = 0$	1	2	3	4	5	6	7	8	9	10
$p = 0·01$	0·8345	0·9862	0·9993								
·02	·6951	·9505	·9948	·9996							
·03	·5780	·8997	·9843	·9982	·9998						
·04	·4796	·8393	·9667	·9950	·9994	·9999					
0·05	0·3972	0·7735	0·9419	0·9891	0·9985	0·9998					
·06	·3283	·7055	·9102	·9799	·9966	·9995					
·07	·2708	·6378	·8725	·9667	·9933	·9990	0·9999				
·08	·2229	·5719	·8298	·9494	·9884	·9979	·9997				
·09	·1831	·5091	·7832	·9277	·9814	·9962	·9994	0·9999			
0·10	0·1501	0·4503	0·7338	0·9018	0·9718	0·9936	0·9988	0·9998			
·11	·1227	·3958	·6827	·8718	·9595	·9898	·9979	·9997			
·12	·1002	·3460	·6310	·8382	·9442	·9846	·9966	·9994	0·9999		
·13	·0815	·3008	·5794	·8014	·9257	·9778	·9946	·9989	·9998		
·14	·0662	·2602	·5287	·7618	·9041	·9690	·9919	·9983	·9997		
0·15	0·0536	0·2241	0·4797	0·7202	0·8794	0·9581	0·9882	0·9973	0·9995	0·9999	
·16	·0434	·1920	·4327	·6771	·8518	·9449	·9833	·9959	·9992	·9999	
·17	·0349	·1638	·3881	·6331	·8213	·9292	·9771	·9940	·9987	·9998	
·18	·0281	·1391	·3462	·5888	·7884	·9111	·9694	·9914	·9980	·9996	0·9999
·19	·0225	·1176	·3073	·5446	·7533	·8903	·9600	·9880	·9971	·9994	·9999
0·20	0·0180	0·0991	0·2713	0·5010	0·7164	0·8671	0·9487	0·9837	0·9957	0·9991	0·9998
·21	·0144	·0831	·2384	·4586	·6780	·8414	·9355	·9783	·9940	·9986	·9997
·22	·0114	·0694	·2084	·4175	·6387	·8134	·9201	·9717	·9917	·9980	·9996
·23	·0091	·0577	·1813	·3782	·5988	·7832	·9026	·9637	·9888	·9972	·9994
·24	·0072	·0478	·1570	·3409	·5586	·7512	·8829	·9542	·9852	·9961	·9991
0·25	0·0056	0·0395	0·1353	0·3057	0·5187	0·7175	0·8610	0·9431	0·9807	0·9946	0·9988
·26	0·0044	·0324	·1161	·2728	·4792	·6824	·8370	·9301	·9751	·9927	·9982
·27	·0035	·0265	·0991	·2422	·4406	·6462	·8109	·9153	·9684	·9903	·9975
·28	·0027	·0216	·0842	·2140	·4032	·6093	·7829	·8986	·9605	·9873	·9966
·29	·0021	·0176	·0712	·1881	·3671	·5719	·7531	·8800	·9512	·9836	·9954
0·30	0·0016	0·0142	0·0600	0·1646	0·3327	0·5344	0·7217	0·8593	0·9404	0·9790	0·9939
·31	·0013	·0114	·0502	·1432	·2999	·4971	·6889	·8367	·9280	·9736	·9920
·32	·0010	·0092	·0419	·1241	·2691	·4602	·6550	·8122	·9139	·9671	·9896
·33	·0007	·0073	·0348	·1069	·2402	·4241	·6203	·7859	·8981	·9595	·9867
·34	·0006	·0058	·0287	·0917	·2134	·3889	·5849	·7579	·8804	·9506	·9831
0·35	0·0004	0·0046	0·0236	0·0783	0·1886	0·3550	0·5491	0·7283	0·8609	0·9403	0·9788
·36	·0003	·0036	·0193	·0665	·1659	·3224	·5133	·6973	·8396	·9286	·9736
·37	·0002	·0028	·0157	·0561	·1451	·2914	·4776	·6651	·8165	·9153	·9675
·38	·0002	·0022	·0127	·0472	·1263	·2621	·4424	·6319	·7916	·9003	·9603
·39	·0001	·0017	·0103	·0394	·1093	·2345	·4079	·5979	·7650	·8837	·9520
0·40	0·0001	0·0013	0·0082	0·0328	0·0942	0·2088	0·3743	0·5634	0·7368	0·8653	0·9424
·41	·0001	·0010	·0066	·0271	·0807	·1849	·3418	·5287	·7072	·8451	·9314
·42	·0001	·0008	·0052	·0223	·0687	·1628	·3105	·4938	·6764	·8232	·9189
·43		·0006	·0041	·0182	·0582	·1427	·2807	·4592	·6444	·7996	·9049
·44		·0004	·0032	·0148	·0490	·1243	·2524	·4250	·6115	·7742	·8893
0·45		0·0003	0·0025	0·0120	0·0411	0·1077	0·2258	0·3915	0·5778	0·7473	0·8720
·46		·0002	·0019	·0096	·0342	·0928	·2009	·3588	·5438	·7188	·8530
·47		·0002	·0015	·0077	·0283	·0795	·1778	·3272	·5094	·6890	·8323
·48		·0001	·0011	·0061	·0233	·0676	·1564	·2968	·4751	·6579	·8098
·49		·0001	·0009	·0048	·0190	·0572	·1368	·2678	·4409	·6258	·7856
0·50		0·0001	0·0007	0·0038	0·0154	0·0481	0·1189	0·2403	0·4073	0·5927	0·7597

See page 4 for explanation of the use of this table.

TABLE 1. THE BINOMIAL DISTRIBUTION FUNCTION

$n = 18$	$r = 11$	12	13	14	15	16	$n = 19$	$r = 0$	1	2
$p = 0.01$							$p = 0.01$	0.8262	0.9847	0.9991
·02							·02	·6812	·9454	·9939
·03							·03	·5606	·8900	·9817
·04							·04	·4604	·8249	·9616
0·05							0·05	0·3774	0·7547	0·9335
·06							·06	·3086	·6829	·8979
·07							·07	·2519	·6121	·8561
·08							·08	·2051	·5440	·8092
·09							·09	·1666	·4798	·7585
0·10							0·10	0·1351	0·4203	0·7054
·11							·11	·1092	·3658	·6512
·12							·12	·0881	·3165	·5968
·13							·13	·0709	·2723	·5432
·14							·14	·0569	·2331	·4911
0·15							0·15	0·0456	0·1985	0·4413
·16							·16	·0364	·1682	·3941
·17							·17	·0290	·1419	·3500
·18							·18	·0230	·1191	·3090
·19							·19	·0182	·0996	·2713
0·20							0·20	0·0144	0·0829	0·2369
·21							·21	·0113	·0687	·2058
·22	0·9999						·22	·0089	·0566	·1778
·23	·9999						·23	·0070	·0465	·1529
·24	·9998						·24	·0054	·0381	·1308
0·25	0·9998						0·25	0·0042	0·0310	0·1113
·26	·9997	0·9999					·26	·0033	·0251	·0943
·27	·9995	·9999					·27	·0025	·0203	·0795
·28	·9993	·9999					·28	·0019	·0163	·0667
·29	·9990	·9998					·29	·0015	·0131	·0557
0·30	0·9986	0·9997					0·30	0·0011	0·0104	0·0462
·31	·9980	·9996	0·9999				·31	·0009	·0083	·0382
·32	·9973	·9995	·9999				·32	·0007	·0065	·0314
·33	·9964	·9992	·9999				·33	·0005	·0051	·0257
·34	·9953	·9989	·9998				·34	·0004	·0040	·0209
0·35	0·9938	0·9986	0·9997				0·35	0·0003	0·0031	0·0170
·36	·9920	·9981	·9996	0·9999			·36	·0002	·0024	·0137
·37	·9898	·9974	·9995	·9999			·37	·0002	·0019	·0110
·38	·9870	·9966	·9993	·9999			·38	·0001	·0014	·0087
·39	·9837	·9956	·9990	·9998			·39	·0001	·0011	·0069
0·40	0·9797	0·9942	0·9987	0·9998			0·40	0·0001	0·0008	0·0055
·41	·9750	·9926	·9983	·9997			·41		·0006	·0043
·42	·9693	·9906	·9978	·9996	0·9999		·42		·0005	·0033
·43	·9628	·9882	·9971	·9994	·9999		·43		·0004	·0026
·44	·9551	·9853	·9962	·9993	·9999		·44		·0003	·0020
0·45	0·9463	0·9817	0·9951	0·9990	0·9999		0·45		0·0002	0·0015
·46	·9362	·9775	·9937	·9987	·9998		·46		·0001	·0012
·47	·9247	·9725	·9921	·9983	·9997		·47		·0001	·0009
·48	·9117	·9666	·9900	·9977	·9996		·48		·0001	·0007
·49	·8972	·9598	·9875	·9971	·9995	0·9999	·49		·0001	·0005
0·50	0·8811	0·9519	0·9846	0·9962	0·9993	0·9999	0·50			0·0004

See page 4 for explanation of the use of this table.

TABLE 1. THE BINOMIAL DISTRIBUTION FUNCTION

TABLE 1. THE BINOMIAL DISTRIBUTION FUNCTION

$n = 19$	$r = 3$	4	5	6	7	8	9	10	11	12	13
$p = 0{\cdot}01$											
·02	0·9995										
·03	·9978	0·9998									
·04	·9939	·9993	0·9999								
0·05	0·9868	·9980	0·9998								
·06	·9757	·9956	·9994	0·9999							
·07	·9602	·9915	·9986	·9998							
·08	·9398	·9853	·9971	·9996	0·9999						
·09	·9147	·9765	·9949	·9991	·9999						
0·10	0·8850	·9648	0·9914	·9983	·9997						
·11	·8510	·9498	·9865	·9970	·9995	0·9999					
·12	·8133	·9315	·9798	·9952	·9991	·9998					
·13	·7725	·9096	·9710	·9924	·9984	·9997					
·14	·7292	·8842	·9599	·9887	·9974	·9995	0·9999				
0·15	0·6841	·8556	·9463	·9837	·9959	·9992	0·9999				
·16	·6380	·8238	·9300	·9772	·9939	·9986	·9998				
·17	·5915	·7893	·9109	·9690	·9911	·9979	·9996	0·9999			
·18	·5451	·7524	·8890	·9589	·9874	·9968	·9993	·9999			
·19	·4995	·7136	·8643	·9468	·9827	·9953	·9990	·9998			
0·20	0·4551	0·6733	0·8369	0·9324	0·9767	0·9933	0·9984	0·9997			
·21	·4123	·6319	·8071	·9157	·9693	·9907	·9977	·9995	0·9999		
·22	·3715	·5900	·7749	·8966	·9604	·9873	·9966	·9993	·9999		
·23	·3329	·5480	·7408	·8752	·9497	·9831	·9953	·9989	·9998		
·24	·2968	·5064	·7050	·8513	·9371	·9778	·9934	·9984	·9997	0·9999	
0·25	0·2631	0·4654	0·6678	0·8251	0·9225	0·9713	0·9911	0·9977	0·9995	0·9999	
·26	·2320	·4256	·6295	·7968	·9059	·9634	·9881	·9968	·9993	·9999	
·27	·2035	·3871	·5907	·7664	·8871	·9541	·9844	·9956	·9990	·9998	
·28	·1776	·3502	·5516	·7343	·8662	·9432	·9798	·9940	·9985	·9997	
·29	·1542	·3152	·5125	·7005	·8432	·9306	·9742	·9920	·9980	·9996	0·9999
0·30	0·1332	0·2822	0·4739	0·6655	0·8180	0·9161	0·9674	0·9895	0·9972	0·9994	0·9999
·31	·1144	·2514	·4359	·6295	·7909	·8997	·9595	·9863	·9962	·9991	·9998
·32	·0978	·2227	·3990	·5927	·7619	·8814	·9501	·9824	·9949	·9988	·9998
·33	·0831	·1963	·3634	·5555	·7312	·8611	·9392	·9777	·9932	·9983	·9997
·34	·0703	·1720	·3293	·5182	·6990	·8388	·9267	·9720	·9911	·9977	·9995
0·35	0·0591	0·1500	0·2968	0·4812	0·6656	0·8145	0·9125	0·9653	0·9886	0·9969	0·9993
·36	·0495	·1301	·2661	·4446	·6310	·7884	·8965	·9574	·9854	·9959	·9991
·37	·0412	·1122	·2373	·4087	·5957	·7605	·8787	·9482	·9815	·9946	·9987
·38	·0341	·0962	·2105	·3739	·5599	·7309	·8590	·9375	·9769	·9930	·9983
·39	·0281	·0821	·1857	·3403	·5238	·6998	·8374	·9253	·9713	·9909	·9977
0·40	0·0230	0·0696	0·1629	0·3081	0·4878	0·6675	0·8139	0·9115	0·9648	0·9884	0·9969
·41	·0187	·0587	·1421	·2774	·4520	·6340	·7886	·8960	·9571	·9854	·9960
·42	·0151	·0492	·1233	·2485	·4168	·5997	·7615	·8787	·9482	·9817	·9948
·43	·0122	·0410	·1063	·2213	·3824	·5647	·7328	·8596	·9379	·9773	·9933
·44	·0097	·0340	·0912	·1961	·3491	·5294	·7026	·8387	·9262	·9720	·9914
0·45	0·0077	0·0280	0·0777	0·1727	0·3169	0·4940	0·6710	0·8159	0·9129	0·9658	0·9891
·46	·0061	·0229	·0658	·1512	·2862	·4587	·6383	·7913	·8979	·9585	·9863
·47	·0048	·0186	·0554	·1316	·2570	·4238	·6046	·7649	·8813	·9500	·9829
·48	·0037	·0150	·0463	·1138	·2294	·3895	·5701	·7369	·8628	·9403	·9788
·49	·0029	·0121	·0385	·0978	·2036	·3561	·5352	·7072	·8425	·9291	·9739
0·50	0·0022	0·0096	0·0318	0·0835	0·1796	0·3238	0·5000	0·6762	0·8204	0·9165	0·9682

See page 4 for explanation of the use of this table.

TABLE 1. THE BINOMIAL DISTRIBUTION FUNCTION

$n = 19$	$r = 14$	15	16
$p = 0.01$			
.02			
.03			
.04			
0.05			
.06			
.07			
.08			
.09			
0.10			
.11			
.12			
.13			
.14			
0.15			
.16			
.17			
.18			
.19			
0.20			
.21			
.22			
.23			
.24			
0.25			
.26			
.27			
.28			
.29			
0.30			
.31			
.32			
.33	0.9999		
.34	.9999		
0.35	0.9999		
.36	.9998		
.37	.9998		
.38	.9997		
.39	.9995	0.9999	
0.40	0.9994	0.9999	
.41	.9991	.9999	
.42	.9988	.9998	
.43	.9984	.9997	
.44	.9979	.9996	0.9999
0.45	0.9972	0.9995	0.9999
.46	.9964	.9993	.9999
.47	.9954	.9990	.9999
.48	.9940	.9987	.9998
.49	.9924	.9983	.9997
0.50	0.9904	0.9978	0.9996

$n = 20$	$r = 0$	1	2	3	4	5	6
$p = 0.01$	0.8179	0.9831	0.9990				
.02	.6676	.9401	.9929	.9994			
.03	.5438	.8802	.9790	.9973	0.9997		
.04	.4420	.8103	.9561	.9926	.9990	0.9999	
0.05	0.3585	0.7358	0.9245	0.9841	0.9974	0.9997	
.06	.2901	.6605	.8850	.9710	.9944	.9991	0.9999
.07	.2342	.5869	.8390	.9529	.9893	.9981	.9997
.08	.1887	.5169	.7879	.9294	.9817	.9962	.9994
.09	.1516	.4516	.7334	.9007	.9710	.9932	.9987
0.10	0.1216	0.3917	0.6769	0.8670	0.9568	0.9887	0.9976
.11	.0972	.3376	.6198	.8290	.9390	.9825	.9959
.12	.0776	.2891	.5631	.7873	.9173	.9740	.9933
.13	.0617	.2461	.5080	.7427	.8917	.9630	.9897
.14	.0490	.2084	.4550	.6959	.8625	.9493	.9847
0.15	0.0388	0.1756	0.4049	0.6477	0.8298	0.9327	0.9781
.16	.0306	.1471	.3580	.5990	.7941	.9130	.9696
.17	.0241	.1227	.3146	.5504	.7557	.8902	.9591
.18	.0189	.1018	.2748	.5026	.7151	.8644	.9463
.19	.0148	.0841	.2386	.4561	.6729	.8357	.9311
0.20	0.0115	0.0692	0.2061	0.4114	0.6296	0.8042	0.9133
.21	.0090	.0566	.1770	.3690	.5858	.7703	.8929
.22	.0069	.0461	.1512	.3289	.5420	.7343	.8699
.23	.0054	.0374	.1284	.2915	.4986	.6965	.8443
.24	.0041	.0302	.1085	.2569	.4561	.6573	.8162
0.25	0.0032	0.0243	0.0913	0.2252	0.4148	0.6172	0.7858
.26	.0024	.0195	.0763	.1962	.3752	.5765	.7533
.27	.0018	.0155	.0635	.1700	.3375	.5357	.7190
.28	.0014	.0123	.0526	.1466	.3019	.4952	.6831
.29	.0011	.0097	.0433	.1256	.2685	.4553	.6460
0.30	0.0008	0.0076	0.0355	0.1071	0.2375	0.4164	0.6080
.31	.0006	.0060	.0289	.0908	.2089	.3787	.5695
.32	.0004	.0047	.0235	.0765	.1827	.3426	.5307
.33	.0003	.0036	.0189	.0642	.1589	.3083	.4921
.34	.0002	.0028	.0152	.0535	.1374	.2758	.4540
0.35	0.0002	0.0021	0.0121	0.0444	0.1182	0.2454	0.4166
.36	.0001	.0016	.0096	.0366	.1011	.2171	.3803
.37	.0001	.0012	.0076	.0300	.0859	.1910	.3453
.38	.0001	.0009	.0060	.0245	.0726	.1671	.3118
.39	.0001	.0007	.0047	.0198	.0610	.1453	.2800
0.40		0.0005	0.0036	0.0160	0.0510	0.1256	0.2500
.41		.0004	.0028	.0128	.0423	.1079	.2220
.42		.0003	.0021	.0102	.0349	.0922	.1959
.43		.0002	.0016	.0080	.0286	.0783	.1719
.44		.0002	.0012	.0063	.0233	.0660	.1499
0.45		0.0001	0.0009	0.0049	0.0189	0.0553	0.1299
.46		.0001	.0007	.0038	.0152	.0461	.1119
.47		.0001	.0005	.0029	.0121	.0381	.0958
.48			.0004	.0023	.0096	.0313	.0814
.49			.0003	.0017	.0076	.0255	.0688
0.50			0.0002	0.0013	0.0059	0.0207	0.0577

See page 4 for explanation of the use of this table.

TABLE 1. THE BINOMIAL DISTRIBUTION FUNCTION

$n = 20$	$r = 7$	8	9	10	11	12	13	14	15	16	17
$p = 0\cdot01$											
·02											
·03											
·04											
0·05											
·06											
·07											
·08	0·9999										
·09	·9998										
0·10	0·9996	0·9999									
·11	·9992	·9999									
·12	·9986	·9998									
·13	·9976	·9995	0·9999								
·14	·9962	·9992	·9999								
0·15	0·9941	0·9987	0·9998								
·16	·9912	·9979	·9996	0·9999							
·17	·9873	·9967	·9993	·9999							
·18	·9823	·9951	·9989	·9998							
·19	·9759	·9929	·9983	·9996	0·9999						
0·20	0·9679	0·9900	0·9974	0·9994	0·9999						
·21	·9581	·9862	·9962	·9991	·9998						
·22	·9464	·9814	·9946	·9987	·9997						
·23	·9325	·9754	·9925	·9981	·9996	0·9999					
·24	·9165	·9680	·9897	·9972	·9994	·9999					
0·25	0·8982	0·9591	0·9861	0·9961	0·9991	0·9998					
·26	·8775	·9485	·9817	·9945	·9986	·9997					
·27	·8545	·9360	·9762	·9926	·9981	·9996	0·9999				
·28	·8293	·9216	·9695	·9900	·9973	·9994	·9999				
·29	·8018	·9052	·9615	·9868	·9962	·9991	·9998				
0·30	0·7723	0·8867	0·9520	0·9829	0·9949	0·9987	0·9997				
·31	·7409	·8660	·9409	·9780	·9931	·9982	·9996	0·9999			
·32	·7078	·8432	·9281	·9721	·9909	·9975	·9994	·9999			
·33	·6732	·8182	·9134	·9650	·9881	·9966	·9992	·9999			
·34	·6376	·7913	·8968	·9566	·9846	·9955	·9989	·9998			
0·35	0·6010	0·7624	0·8782	0·9468	0·9804	0·9940	0·9985	0·9997			
·36	·5639	·7317	·8576	·9355	·9753	·9921	·9979	·9996	0·9999		
·37	·5265	·6995	·8350	·9225	·9692	·9898	·9972	·9994	·9999		
·38	·4892	·6659	·8103	·9077	·9619	·9868	·9963	·9991	·9998		
·39	·4522	·6312	·7837	·8910	·9534	·9833	·9951	·9988	·9998		
0·40	0·4159	0·5956	0·7553	0·8725	0·9435	0·9790	0·9935	0·9984	0·9997		
·41	·3804	·5594	·7252	·8520	·9321	·9738	·9916	·9978	·9996	0·9999	
·42	·3461	·5229	·6936	·8295	·9190	·9676	·9893	·9971	·9994	·9999	
·43	·3132	·4864	·6606	·8051	·9042	·9603	·9864	·9962	·9992	·9999	
·44	·2817	·4501	·6264	·7788	·8877	·9518	·9828	·9950	·9989	·9998	
0·45	0·2520	0·4143	0·5914	0·7507	0·8692	0·9420	0·9786	0·9936	0·9985	0·9997	
·46	·2241	·3793	·5557	·7209	·8489	·9306	·9735	·9917	·9980	·9996	0·9999
·47	·1980	·3454	·5196	·6896	·8266	·9177	·9674	·9895	·9973	·9995	·9999
·48	·1739	·3127	·4834	·6568	·8023	·9031	·9603	·9867	·9965	·9993	·9999
·49	·1518	·2814	·4475	·6229	·7762	·8867	·9520	·9834	·9954	·9990	·9999
0·50	0·1316	0·2517	0·4119	0·5881	0·7483	0·8684	0·9423	0·9793	0·9941	0·9987	0·9998

See page 4 for explanation of the use of this table.

TABLE 2. THE POISSON DISTRIBUTION FUNCTION

μ	$r=0$	1	2	3	4	5	6
0·00	1·0000						
·02	0·9802	0·9998					
·04	0·9608	·9992					
·06	0·9418	·9983					
·08	0·9231	·9970	0·9999				
0·10	0·9048	0·9953	0·9998				
·12	·8869	·9934	·9997				
·14	·8694	·9911	·9996				
·16	·8521	·9885	·9994				
·18	·8353	·9856	·9992				
0·20	0·8187	0·9825	0·9989	0·9999			
·22	·8025	·9791	·9985	·9999			
·24	·7866	·9754	·9981	·9999			
·26	·7711	·9715	·9976	·9998			
·28	·7558	·9674	·9970	·9998			
0·30	0·7408	0·9631	0·9964	0·9997			
·32	·7261	·9585	·9957	·9997			
·34	·7118	·9538	·9949	·9996			
·36	·6977	·9488	·9940	·9995			
·38	·6839	·9437	·9931	·9994			
0·40	0·6703	0·9384	0·9921	0·9992	0·9999		
·42	·6570	·9330	·9910	·9991	·9999		
·44	·6440	·9274	·9898	·9989	·9999		
·46	·6313	·9217	·9885	·9987	·9999		
·48	·6188	·9158	·9871	·9985	·9999		
0·50	0·6065	0·9098	0·9856	0·9982	0·9998		
·52	·5945	·9037	·9841	·9980	·9998		
·54	·5827	·8974	·9824	·9977	·9998		
·56	·5712	·8911	·9807	·9974	·9997		
·58	·5599	·8846	·9788	·9970	·9997		
0·60	0·5488	0·8781	0·9769	0·9966	0·9996		
·62	·5379	·8715	·9749	·9962	·9995		
·64	·5273	·8648	·9727	·9958	·9995	0·9999	
·66	·5169	·8580	·9705	·9953	·9994	·9999	
·68	·5066	·8511	·9682	·9948	·9993	·9999	
0·70	0·4966	0·8442	0·9659	0·9942	0·9992	0·9999	
·72	·4868	·8372	·9634	·9937	·9991	·9999	
·74	·4771	·8302	·9608	·9930	·9990	·9999	
·76	·4677	·8231	·9582	·9924	·9989	·9999	
·78	·4584	·8160	·9554	·9917	·9987	·9998	
0·80	0·4493	0·8088	0·9526	0·9909	0·9986	0·9998	
·82	·4404	·8016	·9497	·9901	·9984	·9998	
·84	·4317	·7943	·9467	·9893	·9983	·9998	
·86	·4232	·7871	·9436	·9884	·9981	·9997	
·88	·4148	·7798	·9404	·9875	·9979	·9997	
0·90	0·4066	0·7725	0·9371	0·9865	0·9977	0·9997	
·92	·3985	·7652	·9338	·9855	·9974	·9996	
·94	·3906	·7578	·9304	·9845	·9972	·9996	0·9999
·96	·3829	·7505	·9269	·9834	·9969	·9995	·9999
·98	·3753	·7431	·9233	·9822	·9966	·9995	·9999
1·00	0·3679	0·7358	0·9197	0·9810	0·9963	0·9994	0·9999

The function tabulated is

$$F(r|\mu) = \sum_{t=0}^{r} e^{-\mu} \frac{\mu^t}{t!}$$

for $r = 0, 1, 2, \ldots$ and $\mu \leqslant 20$. If R is a random variable with a Poisson distribution with mean μ, $F(r|\mu)$ is the probability that $R \leqslant r$; that is,

$$\Pr\{R \leqslant r\} = F(r|\mu).$$

Note that

$$\Pr\{R \geqslant r\} = 1 - \Pr\{R \leqslant r-1\}$$
$$= 1 - F(r-1|\mu).$$

The probability of *exactly* r occurrences, $\Pr\{R = r\}$, is equal to

$$F(r|\mu) - F(r-1|\mu) = e^{-\mu}\frac{\mu^r}{r!}.$$

Linear interpolation in μ is satisfactory over much of the table, but there are places where quadratic interpolation is necessary for high accuracy. Even quadratic interpolation may be unsatisfactory when $r = 0$ or 1 and a direct calculation is to be preferred: $F(0|\mu) = e^{-\mu}$ and $F(1|\mu) = e^{-\mu}(1 + \mu)$.

For $\mu > 20$, R is approximately normally distributed with mean μ and variance μ; hence, including $\frac{1}{2}$ for continuity, we have

$$F(r|\mu) \doteqdot \Phi(s)$$

where $s = (r + \frac{1}{2} - \mu)/\sqrt{\mu}$ and $\Phi(s)$ is the normal distribution function (see Table 4). The approximation can usually be improved by using the formula

$$F(r|\mu) \doteqdot \Phi(s) - \frac{1}{\sqrt{2\pi}} e^{-\frac{1}{2}s^2} \left\{ \frac{(s^2 - 1)}{6\sqrt{\mu}} + \frac{(s^5 - 7s^3 + 6s)}{72\mu} \right\}.$$

For certain values of r and $\mu > 20$ use may be made of the following relation between the Poisson and χ^2-distributions:

$$F(r|\mu) = 1 - F_{2(r+1)}(2\mu)$$

where $F_\nu(x)$ is the χ^2-distribution function (see Table 7).

Omitted entries to the left and right of tabulated values are 0 and 1 respectively, to four decimal places.

TABLE 2. THE POISSON DISTRIBUTION FUNCTION

μ	r = 0	1	2	3	4	5	6	7	8	9	10	11
1.00	0.3679	0.7358	0.9197	0.9810	0.9963	0.9994	0.9999					
.05	.3499	.7174	.9103	.9778	.9955	.9992	.9999					
.10	.3329	.6990	.9004	.9743	.9946	.9990	.9999					
.15	.3166	.6808	.8901	.9704	.9935	.9988	.9998					
.20	.3012	.6626	.8795	.9662	.9923	.9985	.9997					
1.25	0.2865	0.6446	0.8685	0.9617	0.9909	0.9982	0.9997					
.30	.2725	.6268	.8571	.9569	.9893	.9978	.9996	0.9999				
.35	.2592	.6092	.8454	.9518	.9876	.9973	.9995	.9999				
.40	.2466	.5918	.8335	.9463	.9857	.9968	.9994	.9999				
.45	.2346	.5747	.8213	.9405	.9837	.9962	.9992	.9999				
1.50	0.2231	0.5578	0.8088	0.9344	0.9814	0.9955	0.9991	0.9998				
.55	.2122	.5412	.7962	.9279	.9790	.9948	.9989	.9998				
.60	.2019	.5249	.7834	.9212	.9763	.9940	.9987	.9997				
.65	.1920	.5089	.7704	.9141	.9735	.9930	.9984	.9997	0.9999			
.70	.1827	.4932	.7572	.9068	.9704	.9920	.9981	.9996	.9999			
1.75	0.1738	0.4779	0.7440	0.8992	0.9671	0.9909	0.9978	0.9995	0.9999			
.80	.1653	.4628	.7306	.8913	.9636	.9896	.9974	.9994	.9999			
.85	.1572	.4481	.7172	.8831	.9599	.9883	.9970	.9993	.9999			
.90	.1496	.4337	.7037	.8747	.9559	.9868	.9966	.9992	.9998			
.95	.1423	.4197	.6902	.8660	.9517	.9852	.9960	.9991	.9998			
2.00	0.1353	0.4060	0.6767	0.8571	0.9473	0.9834	0.9955	0.9989	0.9998			
.05	.1287	.3926	.6631	.8480	.9427	.9816	.9948	.9987	.9997	0.9999		
.10	.1225	.3796	.6496	.8386	.9379	.9796	.9941	.9985	.9997	.9999		
.15	.1165	.3669	.6361	.8291	.9328	.9774	.9934	.9983	.9996	.9999		
.20	.1108	.3546	.6227	.8194	.9275	.9751	.9925	.9980	.9995	.9999		
2.25	0.1054	0.3425	0.6093	0.8094	0.9220	0.9726	0.9916	0.9977	0.9994	0.9999		
.30	.1003	.3309	.5960	.7993	.9162	.9700	.9906	.9974	.9994	.9999		
.35	.0954	.3195	.5828	.7891	.9103	.9673	.9896	.9971	.9993	.9998		
.40	.0907	.3084	.5697	.7787	.9041	.9643	.9884	.9967	.9991	.9998		
.45	.0863	.2977	.5567	.7682	.8978	.9612	.9872	.9962	.9990	.9998	0.9999	
2.50	0.0821	0.2873	0.5438	0.7576	0.8912	0.9580	0.9858	0.9958	0.9989	0.9997	0.9999	
.55	.0781	.2772	.5311	.7468	.8844	.9546	.9844	.9952	.9987	.9997	.9999	
.60	.0743	.2674	.5184	.7360	.8774	.9510	.9828	.9947	.9985	.9996	.9999	
.65	.0707	.2579	.5060	.7251	.8703	.9472	.9812	.9940	.9983	.9996	.9999	
.70	.0672	.2487	.4936	.7141	.8629	.9433	.9794	.9934	.9981	.9995	.9999	
2.75	0.0639	0.2397	0.4815	0.7030	0.8554	0.9392	0.9776	0.9927	0.9978	0.9994	0.9999	
.80	.0608	.2311	.4695	.6919	.8477	.9349	.9756	.9919	.9976	.9993	.9998	
.85	.0578	.2227	.4576	.6808	.8398	.9304	.9735	.9910	.9973	.9992	.9998	
.90	.0550	.2146	.4460	.6696	.8318	.9258	.9713	.9901	.9969	.9991	.9998	0.9999
.95	.0523	.2067	.4345	.6584	.8236	.9210	.9689	.9891	.9966	.9990	.9997	.9999
3.00	0.0498	0.1991	0.4232	0.6472	0.8153	0.9161	0.9665	0.9881	0.9962	0.9989	0.9997	0.9999
.05	.0474	.1918	.4121	.6360	.8068	.9110	.9639	.9870	.9958	.9988	.9997	.9999
.10	.0450	.1847	.4012	.6248	.7982	.9057	.9612	.9858	.9953	.9986	.9996	.9999
.15	.0429	.1778	.3904	.6137	.7895	.9002	.9584	.9845	.9948	.9984	.9996	.9999
.20	.0408	.1712	.3799	.6025	.7806	.8946	.9554	.9832	.9943	.9982	.9995	.9999
3.25	0.0388	0.1648	0.3696	0.5914	0.7717	0.8888	0.9523	0.9817	0.9937	0.9980	0.9994	0.9999
.30	.0369	.1586	.3594	.5803	.7626	.8829	.9490	.9802	.9931	.9978	.9994	.9998
.35	.0351	.1526	.3495	.5693	.7534	.8768	.9457	.9786	.9924	.9976	.9993	.9998
.40	.0334	.1468	.3397	.5584	.7442	.8705	.9421	.9769	.9917	.9973	.9992	.9998
.45	.0317	.1413	.3302	.5475	.7349	.8642	.9385	.9751	.9909	.9970	.9991	.9997
3.50	0.0302	0.1359	0.3208	0.5366	0.7254	0.8576	0.9347	0.9733	0.9901	0.9967	0.9990	0.9997

μ	r = 12
3.40	0.9999
.45	.9999
3.50	0.9999

See page 24 for explanation of the use of this table.

TABLE 2. THE POISSON DISTRIBUTION FUNCTION

μ	r = 0	1	2	3	4	5	6	7	8	9	10
3·50	0·0302	0·1359	0·3208	0·5366	0·7254	0·8576	0·9347	0·9733	0·9901	0·9967	0·9990
·55	·0287	·1307	·3117	·5259	·7160	·8509	·9308	·9713	·9893	·9963	·9989
·60	·0273	·1257	·3027	·5152	·7064	·8441	·9267	·9692	·9883	·9960	·9987
·65	·0260	·1209	·2940	·5046	·6969	·8372	·9225	·9670	·9873	·9956	·9986
·70	·0247	·1162	·2854	·4942	·6872	·8301	·9182	·9648	·9863	·9952	·9984
3·75	0·0235	0·1117	0·2771	0·4838	0·6775	0·8229	0·9137	0·9624	0·9852	0·9947	0·9983
·80	·0224	·1074	·2689	·4735	·6678	·8156	·9091	·9599	·9840	·9942	·9981
·85	·0213	·1032	·2609	·4633	·6581	·8081	·9044	·9573	·9828	·9937	·9979
·90	·0202	·0992	·2531	·4532	·6484	·8006	·8995	·9546	·9815	·9931	·9977
·95	·0193	·0953	·2455	·4433	·6386	·7929	·8945	·9518	·9801	·9925	·9974
4·00	0·0183	0·0916	0·2381	0·4335	0·6288	0·7851	0·8893	0·9489	0·9786	0·9919	0·9972
·05	·0174	·0880	·2309	·4238	·6191	·7773	·8841	·9458	·9771	·9912	·9969
·10	·0166	·0845	·2238	·4142	·6093	·7693	·8786	·9427	·9755	·9905	·9966
·15	·0158	·0812	·2169	·4047	·5996	·7613	·8731	·9394	·9738	·9897	·9963
·20	·0150	·0780	·2102	·3954	·5898	·7531	·8675	·9361	·9721	·9889	·9959
4·25	0·0143	0·0749	0·2037	0·3862	0·5801	0·7449	0·8617	0·9326	0·9702	0·9880	0·9956
·30	·0136	·0719	·1974	·3772	·5704	·7367	·8558	·9290	·9683	·9871	·9952
·35	·0129	·0691	·1912	·3682	·5608	·7283	·8498	·9253	·9663	·9861	·9948
·40	·0123	·0663	·1851	·3594	·5512	·7199	·8436	·9214	·9642	·9851	·9943
·45	·0117	·0636	·1793	·3508	·5416	·7114	·8374	·9175	·9620	·9840	·9938
4·50	0·0111	0·0611	0·1736	0·3423	0·5321	0·7029	0·8311	0·9134	0·9597	0·9829	0·9933
·55	·0106	·0586	·1680	·3339	·5226	·6944	·8246	·9092	·9574	·9817	·9928
·60	·0101	·0563	·1626	·3257	·5132	·6858	·8180	·9049	·9549	·9805	·9922
·65	·0096	·0540	·1574	·3176	·5039	·6771	·8114	·9005	·9524	·9792	·9916
·70	·0091	·0518	·1523	·3097	·4946	·6684	·8046	·8960	·9497	·9778	·9910
4·75	0·0087	0·0497	0·1473	0·3019	0·4854	0·6597	0·7978	0·8914	0·9470	0·9764	0·9903
·80	·0082	·0477	·1425	·2942	·4763	·6510	·7908	·8867	·9442	·9749	·9896
·85	·0078	·0458	·1379	·2867	·4672	·6423	·7838	·8818	·9413	·9733	·9888
·90	·0074	·0439	·1333	·2793	·4582	·6335	·7767	·8769	·9382	·9717	·9880
·95	·0071	·0421	·1289	·2721	·4493	·6247	·7695	·8718	·9351	·9699	·9872
5·00	0·0067	0·0404	0·1247	0·2650	0·4405	0·6160	0·7622	0·8666	0·9319	0·9682	0·9863
·05	·0064	·0388	·1205	·2581	·4318	·6072	·7548	·8614	·9286	·9663	·9854
·10	·0061	·0372	·1165	·2513	·4231	·5984	·7474	·8560	·9252	·9644	·9844
·15	·0058	·0357	·1126	·2446	·4146	·5897	·7399	·8505	·9217	·9624	·9834
·20	·0055	·0342	·1088	·2381	·4061	·5809	·7324	·8449	·9181	·9603	·9823
5·25	0·0052	0·0328	0·1051	0·2317	0·3978	0·5722	0·7248	0·8392	0·9144	0·9582	0·9812
·30	·0050	·0314	·1016	·2254	·3895	·5635	·7171	·8335	·9106	·9559	·9800
·35	·0047	·0302	·0981	·2193	·3814	·5548	·7094	·8276	·9067	·9536	·9788
·40	·0045	·0289	·0948	·2133	·3733	·5461	·7017	·8217	·9027	·9512	·9775
·45	·0043	·0277	·0915	·2074	·3654	·5375	·6939	·8156	·8986	·9488	·9761
5·50	0·0041	0·0266	0·0884	0·2017	0·3575	0·5289	0·6860	0·8095	0·8944	0·9462	0·9747
·55	·0039	·0255	·0853	·1961	·3498	·5204	·6782	·8033	·8901	·9436	·9733
·60	·0037	·0244	·0824	·1906	·3422	·5119	·6703	·7970	·8857	·9409	·9718
·65	·0035	·0234	·0795	·1853	·3346	·5034	·6623	·7906	·8812	·9381	·9702
·70	·0033	·0224	·0768	·1800	·3272	·4950	·6544	·7841	·8766	·9352	·9686
5·75	0·0032	0·0215	0·0741	0·1749	0·3199	0·4866	0·6464	0·7776	0·8719	0·9322	0·9669
·80	·0030	·0206	·0715	·1700	·3127	·4783	·6384	·7710	·8672	·9292	·9651
·85	·0029	·0197	·0690	·1651	·3056	·4701	·6304	·7644	·8623	·9260	·9633
·90	·0027	·0189	·0666	·1604	·2987	·4619	·6224	·7576	·8574	·9228	·9614
·95	·0026	·0181	·0642	·1557	·2918	·4537	·6143	·7508	·8524	·9195	·9594
6·00	0·0025	0·0174	0·0620	0·1512	0·2851	0·4457	0·6063	0·7440	0·8472	0·9161	0·9574

See page 24 for explanation of the use of this table.

TABLE 2. THE POISSON DISTRIBUTION FUNCTION

μ	r = 11	12	13	14	15	16	17
3·50	0·9997	0·9999					
·55	·9997	·9999					
·60	·9996	·9999					
·65	·9996	·9999					
·70	·9995	·9999					
3·75	0·9995	0·9999					
·80	·9994	·9998					
·85	·9993	·9998	0·9999				
·90	·9993	·9998	·9999				
·95	·9992	·9998	·9999				
4·00	0·9991	0·9997	0·9999				
·05	·9990	·9997	·9999				
·10	·9989	·9997	·9999				
·15	·9988	·9996	·9999				
·20	·9986	·9996	·9999				
4·25	0·9985	0·9995	0·9999				
·30	·9983	·9995	·9998				
·35	·9982	·9994	·9998	0·9999			
·40	·9980	·9993	·9998	·9999			
·45	·9978	·9993	·9998	·9999			
4·50	0·9976	0·9992	0·9997	0·9999			
·55	·9974	·9991	·9997	·9999			
·60	·9971	·9990	·9997	·9999			
·65	·9969	·9989	·9997	·9999			
·70	·9966	·9988	·9996	·9999			
4·75	0·9963	0·9987	0·9996	0·9999			
·80	·9960	·9986	·9995	·9999			
·85	·9957	·9984	·9995	·9998			
·90	·9953	·9983	·9994	·9998	0·9999		
·95	·9949	·9981	·9994	·9998	·9999		
5·00	0·9945	0·9980	0·9993	0·9998	0·9999		
·05	·9941	·9978	·9992	·9997	·9999		
·10	·9937	·9976	·9992	·9997	·9999		
·15	·9932	·9974	·9991	·9997	·9999		
·20	·9927	·9972	·9990	·9997	·9999		
5·25	0·9922	0·9970	0·9989	0·9996	0·9999		
·30	·9916	·9967	·9988	·9996	·9999		
·35	·9910	·9964	·9987	·9995	·9999		
·40	·9904	·9962	·9986	·9995	·9998	0·9999	
·45	·9897	·9959	·9984	·9995	·9998	·9999	
5·50	0·9890	0·9955	0·9983	0·9994	0·9998	0·9999	
·55	·9883	·9952	·9982	·9993	·9998	·9999	
·60	·9875	·9949	·9980	·9993	·9998	·9999	
·65	·9867	·9945	·9979	·9992	·9997	·9999	
·70	·9859	·9941	·9977	·9991	·9997	·9999	
5·75	0·9850	0·9937	0·9975	0·9991	0·9997	0·9999	
·80	·9841	·9932	·9973	·9990	·9996	·9999	
·85	·9831	·9927	·9971	·9989	·9996	·9999	
·90	·9821	·9922	·9969	·9988	·9996	·9999	
·95	·9810	·9917	·9966	·9987	·9995	·9998	0·9999
6·00	0·9799	0·9912	0·9964	0·9986	0·9995	0·9998	0·9999

μ	r = 0	1	2
6·0	0·0025	0·0174	0·0620
·1	·0022	·0159	·0577
·2	·0020	·0146	·0536
·3	·0018	·0134	·0498
·4	·0017	·0123	·0463
6·5	0·0015	0·0113	0·0430
·6	·0014	·0103	·0400
·7	·0012	·0095	·0371
·8	·0011	·0087	·0344
·9	·0010	·0080	·0320
7·0	0·0009	0·0073	0·0296
·1	·0008	·0067	·0275
·2	·0007	·0061	·0255
·3	·0007	·0056	·0236
·4	·0006	·0051	·0219
7·5	0·0006	0·0047	0·0203
·6	·0005	·0043	·0188
·7	·0005	·0039	·0174
·8	·0004	·0036	·0161
·9	·0004	·0033	·0149
8·0	0·0003	0·0030	0·0138
·1	·0003	·0028	·0127
·2	·0003	·0025	·0118
·3	·0002	·0023	·0109
·4	·0002	·0021	·0100
8·5	0·0002	0·0019	0·0093
·6	·0002	·0018	·0086
·7	·0002	·0016	·0079
·8	·0002	·0015	·0073
·9	·0001	·0014	·0068
9·0	0·0001	0·0012	0·0062
·1	·0001	·0011	·0058
·2	·0001	·0010	·0053
·3	·0001	·0009	·0049
·4	·0001	·0009	·0045
9·5	0·0001	0·0008	0·0042
·6	·0001	·0007	·0038
·7	·0001	·0007	·0035
·8	·0001	·0006	·0033
·9	·0001	·0005	·0030
10·0		0·0005	0·0028
·1		·0005	·0026
·2		·0004	·0023
·3		·0004	·0022
·4		·0003	·0020
10·5		0·0003	0·0018
·6		·0003	·0017
·7		·0003	·0016
·8		·0002	·0014
·9		·0002	·0013
11·0		0·0002	0·0012

See page 24 for explanation of the use of this table.

TABLE 2. THE POISSON DISTRIBUTION FUNCTION

μ	$r = 3$	4	5	6	7	8	9	10	11	12	13
6·0	0·1512	0·2851	0·4457	0·6063	0·7440	0·8472	0·9161	0·9574	0·9799	0·9912	0·9964
·1	·1425	·2719	·4298	·5902	·7301	·8367	·9090	·9531	·9776	·9900	·9958
·2	·1342	·2592	·4141	·5742	·7160	·8259	·9016	·9486	·9750	·9887	·9952
·3	·1264	·2469	·3988	·5582	·7017	·8148	·8939	·9437	·9723	·9873	·9945
·4	·1189	·2351	·3837	·5423	·6873	·8033	·8858	·9386	·9693	·9857	·9937
6·5	0·1118	0·2237	0·3690	0·5265	0·6728	0·7916	0·8774	0·9332	0·9661	0·9840	0·9929
·6	·1052	·2127	·3547	·5108	·6581	·7796	·8686	·9274	·9627	·9821	·9920
·7	·0988	·2022	·3406	·4953	·6433	·7673	·8596	·9214	·9591	·9801	·9909
·8	·0928	·1920	·3270	·4799	·6285	·7548	·8502	·9151	·9552	·9779	·9898
·9	·0871	·1823	·3137	·4647	·6136	·7420	·8405	·9084	·9510	·9755	·9885
7·0	0·0818	0·1730	0·3007	0·4497	0·5987	0·7291	0·8305	0·9015	0·9467	0·9730	0·9872
·1	·0767	·1641	·2881	·4349	·5838	·7160	·8202	·8942	·9420	·9703	·9857
·2	·0719	·1555	·2759	·4204	·5689	·7027	·8096	·8867	·9371	·9673	·9841
·3	·0674	·1473	·2640	·4060	·5541	·6892	·7988	·8788	·9319	·9642	·9824
·4	·0632	·1395	·2526	·3920	·5393	·6757	·7877	·8707	·9265	·9609	·9805
7·5	0·0591	0·1321	0·2414	0·3782	0·5246	0·6620	0·7764	0·8622	0·9208	0·9573	0·9784
·6	·0554	·1249	·2307	·3646	·5100	·6482	·7649	·8535	·9148	·9536	·9762
·7	·0518	·1181	·2203	·3514	·4956	·6343	·7531	·8445	·9085	·9496	·9739
·8	·0485	·1117	·2103	·3384	·4812	·6204	·7411	·8352	·9020	·9454	·9714
·9	·0453	·1055	·2006	·3257	·4670	·6065	·7290	·8257	·8952	·9409	·9687
8·0	0·0424	0·0996	0·1912	0·3134	0·4530	0·5925	0·7166	0·8159	0·8881	0·9362	0·9658
·1	·0396	·0940	·1822	·3013	·4391	·5786	·7041	·8058	·8807	·9313	·9628
·2	·0370	·0887	·1736	·2896	·4254	·5647	·6915	·7955	·8731	·9261	·9595
·3	·0346	·0837	·1653	·2781	·4119	·5507	·6788	·7850	·8652	·9207	·9561
·4	·0323	·0789	·1573	·2670	·3987	·5369	·6659	·7743	·8571	·9150	·9524
8·5	0·0301	0·0744	0·1496	0·2562	0·3856	0·5231	0·6530	0·7634	0·8487	0·9091	0·9486
·6	·0281	·0701	·1422	·2457	·3728	·5094	·6400	·7522	·8400	·9029	·9445
·7	·0262	·0660	·1352	·2355	·3602	·4958	·6269	·7409	·8311	·8965	·9403
·8	·0244	·0621	·1284	·2256	·3478	·4823	·6137	·7294	·8220	·8898	·9358
·9	·0228	·0584	·1219	·2160	·3357	·4689	·6006	·7178	·8126	·8829	·9311
9·0	0·0212	0·0550	0·1157	0·2068	0·3239	0·4557	0·5874	0·7060	0·8030	0·8758	0·9261
·1	·0198	·0517	·1098	·1978	·3123	·4426	·5742	·6941	·7932	·8684	·9210
·2	·0184	·0486	·1041	·1892	·3010	·4296	·5611	·6820	·7832	·8607	·9156
·3	·0172	·0456	·0986	·1808	·2900	·4168	·5479	·6699	·7730	·8529	·9100
·4	·0160	·0429	·0935	·1727	·2792	·4042	·5349	·6576	·7626	·8448	·9042
9·5	0·0149	0·0403	0·0885	0·1649	0·2687	0·3918	0·5218	0·6453	0·7520	0·8364	0·8981
·6	·0138	·0378	·0838	·1574	·2584	·3796	·5089	·6329	·7412	·8279	·8919
·7	·0129	·0355	·0793	·1502	·2485	·3676	·4960	·6205	·7303	·8191	·8853
·8	·0120	·0333	·0750	·1433	·2388	·3558	·4832	·6080	·7193	·8101	·8786
·9	·0111	·0312	·0710	·1366	·2294	·3442	·4705	·5955	·7081	·8009	·8716
10·0	0·0103	0·0293	0·0671	0·1301	0·2202	0·3328	0·4579	0·5830	0·6968	0·7916	0·8645
·1	·0096	·0274	·0634	·1240	·2113	·3217	·4455	·5705	·6853	·7820	·8571
·2	·0089	·0257	·0599	·1180	·2027	·3108	·4332	·5580	·6738	·7722	·8494
·3	·0083	·0241	·0566	·1123	·1944	·3001	·4210	·5456	·6622	·7623	·8416
·4	·0077	·0225	·0534	·1069	·1863	·2896	·4090	·5331	·6505	·7522	·8336
10·5	0·0071	0·0211	0·0504	0·1016	0·1785	0·2794	0·3971	0·5207	0·6387	0·7420	0·8253
·6	·0066	·0197	·0475	·0966	·1710	·2694	·3854	·5084	·6269	·7316	·8169
·7	·0062	·0185	·0448	·0918	·1636	·2597	·3739	·4961	·6150	·7210	·8083
·8	·0057	·0173	·0423	·0872	·1566	·2502	·3626	·4840	·6031	·7104	·7995
·9	·0053	·0162	·0398	·0828	·1498	·2410	·3515	·4719	·5912	·6996	·7905
11·0	0·0049	0·0151	0·0375	0·0786	0·1432	0·2320	0·3405	0·4599	0·5793	0·6887	0·7813

See page 24 for explanation of the use of this table.

TABLE 2. THE POISSON DISTRIBUTION FUNCTION

μ	$r=14$	15	16	17	18	19	20	21	22	23	24
6·0	0·9986	0·9995	0·9998	0·9999							
·1	·9984	·9994	·9998	·9999							
·2	·9981	·9993	·9997	·9999							
·3	·9978	·9992	·9997	·9999							
·4	·9974	·9990	·9996	·9999							
6·5	0·9970	0·9988	0·9996	0·9998	0·9999						
·6	·9966	·9986	·9995	·9998	·9999						
·7	·9961	·9984	·9994	·9998	·9999						
·8	·9956	·9982	·9993	·9997	·9999						
·9	·9950	·9979	·9992	·9997	·9999						
7·0	0·9943	0·9976	0·9990	0·9996	0·9999						
·1	·9935	·9972	·9989	·9996	·9998	0·9999					
·2	·9927	·9969	·9987	·9995	·9998	·9999					
·3	·9918	·9964	·9985	·9994	·9998	·9999					
·4	·9908	·9959	·9983	·9993	·9997	·9999					
7·5	0·9897	0·9954	0·9980	0·9992	0·9997	0·9999					
·6	·9886	·9948	·9978	·9991	·9996	·9999					
·7	·9873	·9941	·9974	·9989	·9996	·9998	0·9999				
·8	·9859	·9934	·9971	·9988	·9995	·9998	·9999				
·9	·9844	·9926	·9967	·9986	·9994	·9998	·9999				
8·0	0·9827	0·9918	0·9963	0·9984	0·9993	0·9997	0·9999				
·1	·9810	·9908	·9958	·9982	·9992	·9997	·9999				
·2	·9791	·9898	·9953	·9979	·9991	·9997	·9999				
·3	·9771	·9887	·9947	·9977	·9990	·9996	·9998	0·9999			
·4	·9749	·9875	·9941	·9973	·9989	·9995	·9998	·9999			
8·5	0·9726	0·9862	0·9934	0·9970	0·9987	0·9995	0·9998	0·9999			
·6	·9701	·9848	·9926	·9966	·9985	·9994	·9998	·9999			
·7	·9675	·9832	·9918	·9962	·9983	·9993	·9997	·9999			
·8	·9647	·9816	·9909	·9957	·9981	·9992	·9997	·9999			
·9	·9617	·9798	·9899	·9952	·9978	·9991	·9996	·9998	0·9999		
9·0	0·9585	0·9780	0·9889	0·9947	0·9976	0·9989	0·9996	0·9998	0·9999		
·1	·9552	·9760	·9878	·9941	·9973	·9988	·9995	·9998	·9999		
·2	·9517	·9738	·9865	·9934	·9969	·9986	·9994	·9998	·9999		
·3	·9480	·9715	·9852	·9927	·9966	·9985	·9993	·9997	·9999		
·4	·9441	·9691	·9838	·9919	·9962	·9983	·9992	·9997	·9999		
9·5	0·9400	0·9665	0·9823	0·9911	0·9957	0·9980	0·9991	0·9996	0·9999	0·9999	
·6	·9357	·9638	·9806	·9902	·9952	·9978	·9990	·9996	·9998	·9999	
·7	·9312	·9609	·9789	·9892	·9947	·9975	·9989	·9995	·9998	·9999	
·8	·9265	·9579	·9770	·9881	·9941	·9972	·9987	·9995	·9998	·9999	
·9	·9216	·9546	·9751	·9870	·9935	·9969	·9986	·9994	·9997	·9999	
10·0	0·9165	0·9513	0·9730	0·9857	0·9928	0·9965	0·9984	0·9993	0·9997	0·9999	
·1	·9112	·9477	·9707	·9844	·9921	·9962	·9982	·9992	·9997	·9999	0·9999
·2	·9057	·9440	·9684	·9830	·9913	·9957	·9980	·9991	·9996	·9998	·9999
·3	·9000	·9400	·9658	·9815	·9904	·9953	·9978	·9990	·9996	·9998	·9999
·4	·8940	·9359	·9632	·9799	·9895	·9948	·9975	·9989	·9995	·9998	·9999
10·5	0·8879	0·9317	0·9604	0·9781	0·9885	0·9942	0·9972	0·9987	0·9994	0·9998	0·9999
·6	·8815	·9272	·9574	·9763	·9874	·9936	·9969	·9986	·9994	·9997	·9999
·7	·8750	·9225	·9543	·9744	·9863	·9930	·9966	·9984	·9993	·9997	·9999
·8	·8682	·9177	·9511	·9723	·9850	·9923	·9962	·9982	·9992	·9996	·9998
·9	·8612	·9126	·9477	·9701	·9837	·9915	·9958	·9980	·9991	·9996	·9998
11·0	0·8540	0·9074	0·9441	0·9678	0·9823	0·9907	0·9953	0·9977	0·9990	0·9995	0·9998

μ	$r=25$
10·7	0·9999
·8	·9999
·9	·9999
11·0	0·9999

See page 24 for explanation of the use of this table.

TABLE 2. THE POISSON DISTRIBUTION FUNCTION

μ	r = 2	3	4	5	6	7	8	9	10	11	12
11·0	0·0012	0·0049	0·0151	0·0375	0·0786	0·1432	0·2320	0·3405	0·4599	0·5793	0·6887
·2	·0010	·0042	·0132	·0333	·0708	·1307	·2147	·3192	·4362	·5554	·6666
·4	·0009	·0036	·0115	·0295	·0636	·1192	·1984	·2987	·4131	·5316	·6442
·6	·0007	·0031	·0100	·0261	·0571	·1085	·1830	·2791	·3905	·5080	·6216
·8	·0006	·0027	·0087	·0230	·0512	·0986	·1686	·2603	·3685	·4847	·5988
12·0	0·0005	0·0023	0·0076	0·0203	0·0458	0·0895	0·1550	0·2424	0·3472	0·4616	0·5760
·2	·0004	·0020	·0066	·0179	·0410	·0811	·1424	·2254	·3266	·4389	·5531
·4	·0004	·0017	·0057	·0158	·0366	·0734	·1305	·2092	·3067	·4167	·5303
·6	·0003	·0014	·0050	·0139	·0326	·0664	·1195	·1939	·2876	·3950	·5077
·8	·0003	·0012	·0043	·0122	·0291	·0599	·1093	·1794	·2693	·3738	·4853
13·0	0·0002	0·0011	0·0037	0·0107	0·0259	0·0540	0·0998	0·1658	0·2517	0·3532	0·4631
·2	·0002	·0009	·0032	·0094	·0230	·0487	·0910	·1530	·2349	·3332	·4413
·4	·0002	·0008	·0028	·0083	·0204	·0438	·0828	·1410	·2189	·3139	·4199
·6	·0001	·0007	·0024	·0072	·0181	·0393	·0753	·1297	·2037	·2952	·3989
·8	·0001	·0006	·0021	·0063	·0161	·0353	·0684	·1192	·1893	·2773	·3784
14·0	0·0001	0·0005	0·0018	0·0055	0·0142	0·0316	0·0621	0·1094	0·1757	0·2600	0·3585
·2	·0001	·0004	·0016	·0048	·0126	·0283	·0562	·1003	·1628	·2435	·3391
·4	·0001	·0003	·0013	·0042	·0111	·0253	·0509	·0918	·1507	·2277	·3203
·6	·0001	·0003	·0012	·0037	·0098	·0226	·0460	·0839	·1392	·2127	·3021
·8		·0002	·0010	·0032	·0087	·0202	·0415	·0766	·1285	·1984	·2845
15·0		0·0002	0·0009	0·0028	0·0076	0·0180	0·0374	0·0699	0·1185	0·1848	0·2676
·2		·0002	·0007	·0024	·0067	·0160	·0337	·0636	·1091	·1718	·2514
·4		·0002	·0006	·0021	·0059	·0143	·0304	·0579	·1003	·1596	·2358
·6		·0001	·0005	·0018	·0052	·0127	·0273	·0526	·0921	·1481	·2209
·8		·0001	·0005	·0016	·0046	·0113	·0245	·0478	·0845	·1372	·2067
16·0		0·0001	0·0004	0·0014	0·0040	0·0100	0·0220	0·0433	0·0774	0·1270	0·1931
·2		·0001	·0003	·0012	·0035	·0089	·0197	·0392	·0708	·1174	·1802
·4		·0001	·0003	·0010	·0031	·0079	·0176	·0355	·0647	·1084	·1680
·6		·0001	·0003	·0009	·0027	·0070	·0158	·0321	·0591	·0999	·1564
·8			·0002	·0008	·0024	·0061	·0141	·0290	·0539	·0920	·1454
17·0			0·0002	0·0007	0·0021	0·0054	0·0126	0·0261	0·0491	0·0847	0·1350
·2			·0002	·0006	·0018	·0048	·0112	·0235	·0447	·0778	·1252
·4			·0001	·0005	·0016	·0042	·0100	·0212	·0406	·0714	·1160
·6			·0001	·0004	·0014	·0037	·0089	·0191	·0369	·0655	·1074
·8			·0001	·0004	·0012	·0033	·0079	·0171	·0335	·0600	·0993
18·0			0·0001	0·0003	0·0010	0·0029	0·0071	0·0154	0·0304	0·0549	0·0917
·2			·0001	·0003	·0009	·0025	·0063	·0138	·0275	·0502	·0846
·4			·0001	·0002	·0008	·0022	·0056	·0124	·0249	·0458	·0779
·6			·0001	·0002	·0007	·0020	·0049	·0111	·0225	·0418	·0717
·8				·0002	·0006	·0017	·0044	·0099	·0203	·0381	·0659
19·0				0·0002	0·0005	0·0015	0·0039	0·0089	0·0183	0·0347	0·0606
·2				·0001	·0005	·0013	·0034	·0079	·0165	·0315	·0556
·4				·0001	·0004	·0012	·0030	·0071	·0149	·0287	·0509
·6				·0001	·0003	·0010	·0027	·0063	·0134	·0260	·0467
·8				·0001	·0003	·0009	·0024	·0056	·0120	·0236	·0427
20·0				0·0001	0·0003	0·0008	0·0021	0·0050	0·0108	0·0214	0·0390

Inset:

μ	r = 1
11·0	0·0002
·2	·0002
·4	·0001
·6	·0001
·8	·0001
12·0	0·0001
·2	·0001
·4	·0001

See page 24 for explanation of the use of this table.

TABLE 2. THE POISSON DISTRIBUTION FUNCTION

μ	$r = 13$	14	15	16	17	18	19	20	21	22	23
11·0	0·7813	0·8540	0·9074	0·9441	0·9678	0·9823	0·9907	0·9953	0·9977	0·9990	0·9995
·2	·7624	·8391	·8963	·9364	·9628	·9792	·9889	·9943	·9972	·9987	·9994
·4	·7430	·8234	·8845	·9280	·9572	·9757	·9868	·9932	·9966	·9984	·9992
·6	·7230	·8069	·8719	·9190	·9511	·9718	·9845	·9918	·9958	·9980	·9991
·8	·7025	·7898	·8585	·9092	·9444	·9674	·9818	·9902	·9950	·9975	·9988
12·0	0·6815	0·7720	0·8444	0·8987	0·9370	0·9626	0·9787	0·9884	0·9939	0·9970	0·9985
·2	·6603	·7536	·8296	·8875	·9290	·9572	·9753	·9863	·9927	·9963	·9982
·4	·6387	·7347	·8140	·8755	·9204	·9513	·9715	·9840	·9914	·9955	·9978
·6	·6169	·7153	·7978	·8629	·9111	·9448	·9672	·9813	·9898	·9946	·9973
·8	·5950	·6954	·7810	·8495	·9011	·9378	·9625	·9783	·9880	·9936	·9967
13·0	0·5730	0·6751	0·7636	0·8355	0·8905	0·9302	0·9573	0·9750	0·9859	0·9924	0·9960
·2	·5511	·6546	·7456	·8208	·8791	·9219	·9516	·9713	·9836	·9910	·9952
·4	·5292	·6338	·7272	·8054	·8671	·9130	·9454	·9671	·9810	·9894	·9943
·6	·5074	·6128	·7083	·7895	·8545	·9035	·9387	·9626	·9780	·9876	·9933
·8	·4858	·5916	·6890	·7730	·8411	·8934	·9314	·9576	·9748	·9856	·9921
14·0	0·4644	0·5704	0·6694	0·7559	0·8272	0·8826	0·9235	0·9521	0·9712	0·9833	0·9907
·2	·4434	·5492	·6494	·7384	·8126	·8712	·9150	·9461	·9671	·9807	·9891
·4	·4227	·5281	·6293	·7204	·7975	·8592	·9060	·9396	·9627	·9779	·9873
·6	·4024	·5071	·6090	·7020	·7818	·8466	·8963	·9326	·9579	·9747	·9853
·8	·3826	·4863	·5886	·6832	·7656	·8333	·8861	·9251	·9526	·9711	·9831
15·0	0·3632	0·4657	0·5681	0·6641	0·7489	0·8195	0·8752	0·9170	0·9469	0·9673	0·9805
·2	·3444	·4453	·5476	·6448	·7317	·8051	·8638	·9084	·9407	·9630	·9777
·4	·3260	·4253	·5272	·6253	·7141	·7901	·8517	·8992	·9340	·9583	·9746
·6	·3083	·4056	·5069	·6056	·6962	·7747	·8391	·8894	·9268	·9532	·9712
·8	·2911	·3864	·4867	·5858	·6779	·7587	·8260	·8791	·9190	·9477	·9674
16·0	0·2745	0·3675	0·4667	0·5660	0·6593	0·7423	0·8122	0·8682	0·9108	0·9418	0·9633
·2	·2585	·3492	·4470	·5461	·6406	·7255	·7980	·8567	·9020	·9353	·9588
·4	·2432	·3313	·4276	·5263	·6216	·7084	·7833	·8447	·8927	·9284	·9539
·6	·2285	·3139	·4085	·5067	·6025	·6908	·7681	·8321	·8828	·9210	·9486
·8	·2144	·2971	·3898	·4871	·5833	·6730	·7524	·8191	·8724	·9131	·9429
17·0	0·2009	0·2808	0·3715	0·4677	0·5640	0·6550	0·7363	0·8055	0·8615	0·9047	0·9367
·2	·1880	·2651	·3535	·4486	·5448	·6367	·7199	·7914	·8500	·8958	·9301
·4	·1758	·2500	·3361	·4297	·5256	·6182	·7031	·7769	·8380	·8864	·9230
·6	·1641	·2354	·3191	·4112	·5065	·5996	·6859	·7619	·8255	·8765	·9154
·8	·1531	·2215	·3026	·3929	·4875	·5810	·6685	·7465	·8126	·8660	·9074
18·0	0·1426	0·2081	0·2867	0·3751	0·4686	0·5622	0·6509	0·7307	0·7991	0·8551	0·8989
·2	·1327	·1953	·2712	·3576	·4500	·5435	·6331	·7146	·7852	·8436	·8899
·4	·1233	·1830	·2563	·3405	·4317	·5249	·6151	·6981	·7709	·8317	·8804
·6	·1145	·1714	·2419	·3239	·4136	·5063	·5970	·6814	·7561	·8193	·8704
·8	·1062	·1603	·2281	·3077	·3958	·4878	·5788	·6644	·7410	·8065	·8600
19·0	0·0984	0·1497	0·2148	0·2920	0·3784	0·4695	0·5606	0·6472	0·7255	0·7931	0·8490
·2	·0911	·1397	·2021	·2768	·3613	·4514	·5424	·6298	·7097	·7794	·8376
·4	·0842	·1303	·1899	·2621	·3446	·4335	·5242	·6122	·6935	·7653	·8257
·6	·0778	·1213	·1782	·2479	·3283	·4158	·5061	·5946	·6772	·7507	·8134
·8	·0717	·1128	·1671	·2342	·3124	·3985	·4881	·5769	·6605	·7358	·8007
20·0	0·0661	0·1049	0·1565	0·2211	0·2970	0·3814	0·4703	0·5591	0·6437	0·7206	0·7875

See page 24 for explanation of the use of this table.

TABLE 2. THE POISSON DISTRIBUTION FUNCTION

μ	r = 35	36	37	38	39
17·2	0·9999				
·4	·9999				
·6	·9999				
·8	·9999				
18·0	0·9999	0·9999			
·2	·9999	·9999			
·4	·9998	·9999			
·6	·9998	·9999	0·9999		
·8	·9997	·9999	·9999		
19·0	0·9997	0·9998	0·9999		
·2	·9996	·9998	·9999		
·4	·9995	·9998	·9999	0·9999	
·6	·9994	·9997	·9999	·9999	
·8	·9993	·9996	·9998	·9999	
20·0	0·9992	0·9996	0·9998	0·9999	0·9999

μ	r = 24	25	26	27	28	29	30	31	32	33	34
11·0	0·9998	0·9999									
·2	·9997	·9999									
·4	·9997	·9999	0·9999								
·6	·9996	·9998	·9999								
·8	·9995	·9998	·9999								
12·0	0·9993	0·9997	0·9999	0·9999							
·2	·9991	·9996	·9998	·9999							
·4	·9989	·9995	·9998	·9999							
·6	·9987	·9994	·9997	·9999	0·9999						
·8	·9984	·9992	·9996	·9998	·9999						
13·0	0·9980	0·9990	0·9995	0·9998	0·9999						
·2	·9976	·9988	·9994	·9997	·9999						
·4	·9971	·9985	·9993	·9997	·9999	0·9999					
·6	·9965	·9982	·9991	·9996	·9998	·9999					
·8	·9958	·9978	·9989	·9995	·9998	·9999					
14·0	0·9950	0·9974	0·9987	0·9994	0·9997	0·9999	0·9999				
·2	·9941	·9969	·9984	·9992	·9996	·9998	·9999				
·4	·9930	·9963	·9981	·9990	·9995	·9998	·9999				
·6	·9918	·9956	·9977	·9988	·9994	·9997	·9999	0·9999			
·8	·9904	·9947	·9972	·9986	·9993	·9997	·9998	·9999			
15·0	0·9888	0·9938	0·9967	0·9983	0·9991	0·9996	0·9998	0·9999			
·2	·9871	·9928	·9961	·9979	·9990	·9995	·9998	·9999	0·9999		
·4	·9851	·9915	·9954	·9975	·9987	·9994	·9997	·9999	·9999		
·6	·9829	·9902	·9945	·9971	·9985	·9992	·9996	·9998	·9999		
·8	·9804	·9886	·9936	·9965	·9982	·9991	·9995	·9998	·9999		
16·0	0·9777	0·9869	0·9925	0·9959	0·9978	0·9989	0·9994	0·9997	0·9999	0·9999	
·2	·9747	·9849	·9913	·9952	·9974	·9986	·9993	·9997	·9998	·9999	
·4	·9713	·9828	·9900	·9944	·9969	·9984	·9992	·9996	·9998	·9999	
·6	·9677	·9804	·9884	·9934	·9964	·9981	·9990	·9995	·9998	·9999	0·9999
·8	·9637	·9777	·9867	·9924	·9957	·9977	·9988	·9994	·9997	·9999	·9999
17·0	0·9594	0·9748	0·9848	0·9912	0·9950	0·9973	0·9986	0·9993	0·9996	0·9998	0·9999
·2	·9546	·9715	·9827	·9898	·9942	·9968	·9983	·9991	·9995	·9998	·9999
·4	·9495	·9680	·9804	·9883	·9933	·9962	·9980	·9989	·9994	·9997	·9999
·6	·9440	·9641	·9778	·9866	·9922	·9956	·9976	·9987	·9993	·9997	·9998
·8	·9381	·9599	·9749	·9848	·9910	·9949	·9972	·9985	·9992	·9996	·9998
18·0	0·9317	0·9554	0·9718	0·9827	0·9897	0·9941	0·9967	0·9982	0·9990	0·9995	0·9998
·2	·9249	·9505	·9683	·9804	·9882	·9931	·9961	·9979	·9989	·9994	·9997
·4	·9177	·9452	·9646	·9779	·9866	·9921	·9955	·9975	·9986	·9993	·9996
·6	·9100	·9395	·9606	·9751	·9847	·9909	·9948	·9971	·9984	·9991	·9996
·8	·9019	·9334	·9562	·9720	·9827	·9896	·9939	·9966	·9981	·9990	·9995
19·0	0·8933	0·9269	0·9514	0·9687	0·9805	0·9882	0·9930	0·9960	0·9978	0·9988	0·9994
·2	·8842	·9199	·9463	·9651	·9780	·9865	·9920	·9954	·9974	·9986	·9992
·4	·8746	·9126	·9409	·9612	·9753	·9847	·9908	·9946	·9970	·9983	·9991
·6	·8646	·9048	·9350	·9570	·9724	·9828	·9895	·9938	·9965	·9980	·9989
·8	·8541	·8965	·9288	·9524	·9692	·9806	·9881	·9929	·9959	·9977	·9987
20·0	0·8432	0·8878	0·9221	0·9475	0·9657	0·9782	0·9865	0·9919	0·9953	0·9973	0·9985

See page 24 for explanation of the use of this table.

TABLE 3. BINOMIAL COEFFICIENTS

This table gives values of

$$\binom{n}{r} = {}^nC_r = \frac{n!}{(n-r)!\,r!} = \frac{n(n-1)\dots(n-r+1)}{r!};$$

when $r > \tfrac{1}{2}n$ use $\binom{n}{r} = \binom{n}{n-r}$. $\binom{n}{r}$ is the number of ways of selecting r objects from n, the order of choice being immaterial. (See also Table 6, which gives values of $\log_{10} n!$ for $n \leqslant 300$.)

r	0	1	2	3	4	5	6	7	8	9
$n = 1$	1	1								
2	1	2	1							
3	1	3	3	1						
4	1	4	6	4	1					
5	1	5	10	10	5	1				
6	1	6	15	20	15	6	1			
7	1	7	21	35	35	21	7	1		
8	1	8	28	56	70	56	28	8	1	
9	1	9	36	84	126	126	84	36	9	1
10	1	10	45	120	210	252	210	120	45	10
11	1	11	55	165	330	462	462	330	165	55
12	1	12	66	220	495	792	924	792	495	220
13	1	13	78	286	715	1287	1716	1716	1287	715
14	1	14	91	364	1001	2002	3003	3432	3003	2002
15	1	15	105	455	1365	3003	5005	6435	6435	5005
16	1	16	120	560	1820	4368	8008	11440	12870	11440
17	1	17	136	680	2380	6188	12376	19448	24310	24310
18	1	18	153	816	3060	8568	18564	31824	43758	48620
19	1	19	171	969	3876	11628	27132	50388	75582	92378
20	1	20	190	1140	4845	15504	38760	77520	125970	167960
21	1	21	210	1330	5985	20349	54264	116280	203490	293930
22	1	22	231	1540	7315	26334	74613	170544	319770	497420
23	1	23	253	1771	8855	33649	100947	245157	490314	817190
24	1	24	276	2024	10626	42504	134596	346104	735471	1307504
25	1	25	300	2300	12650	53130	177100	480700	1081575	2042975
26	1	26	325	2600	14950	65780	230230	657800	1562275	3124550
27	1	27	351	2925	17550	80730	296010	888030	2220075	4686825
28	1	28	378	3276	20475	98280	376740	1184040	3108105	6906900
29	1	29	406	3654	23751	118755	475020	1560780	4292145	10015005
30	1	30	435	4060	27405	142506	593775	2035800	5852925	14307150

r	10	11	12	13	14	15
$n = 20$	184756	167960	125970	77520	38760	15504
21	352716	352716	293930	203490	116280	54264
22	646646	705432	646646	497420	319770	170544
23	1144066	1352078	1352078	1144066	817190	490314
24	1961256	2496144	2704156	2496144	1961256	1307504
25	3268760	4457400	5200300	5200300	4457400	3268760
26	5311735	7726160	9657700	10400600	9657700	7726160
27	8436285	13037895	17383860	20058300	20058300	17383860
28	13123110	21474180	30421755	37442160	40116600	37442160
29	20030010	34597290	51895935	67863915	77558760	77558760
30	30045015	54627300	86493225	119759850	145422675	155117520

TABLE 4. THE NORMAL DISTRIBUTION FUNCTION

The function tabulated is $\Phi(x) = \dfrac{1}{\sqrt{2\pi}} \displaystyle\int_{-\infty}^{x} e^{-\frac{1}{2}t^2}\, dt$. $\Phi(x)$ is

the probability that a random variable, normally distributed with zero mean and unit variance, will be less than or equal to x. When $x < 0$ use $\Phi(x) = 1 - \Phi(-x)$, as the normal distribution with zero mean and unit variance is symmetric about zero.

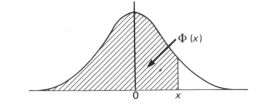

x	$\Phi(x)$	x	$\Phi(x)$	x	$\Phi(x)$	x	$\Phi(x)$	x	$\Phi(x)$	x	$\Phi(x)$
0·00	0·5000	0·40	0·6554	0·80	0·7881	1·20	0·8849	1·60	0·9452	2·00	0·97725
·01	·5040	·41	·6591	·81	·7910	·21	·8869	·61	·9463	·01	·97778
·02	·5080	·42	·6628	·82	·7939	·22	·8888	·62	·9474	·02	·97831
·03	·5120	·43	·6664	·83	·7967	·23	·8907	·63	·9484	·03	·97882
·04	·5160	·44	·6700	·84	·7995	·24	·8925	·64	·9495	·04	·97932
0·05	0·5199	0·45	0·6736	0·85	0·8023	1·25	0·8944	1·65	0·9505	2·05	0·97982
·06	·5239	·46	·6772	·86	·8051	·26	·8962	·66	·9515	·06	·98030
·07	·5279	·47	·6808	·87	·8078	·27	·8980	·67	·9525	·07	·98077
·08	·5319	·48	·6844	·88	·8106	·28	·8997	·68	·9535	·08	·98124
·09	·5359	·49	·6879	·89	·8133	·29	·9015	·69	·9545	·09	·98169
0·10	0·5398	0·50	0·6915	0·90	0·8159	1·30	0·9032	1·70	0·9554	2·10	0·98214
·11	·5438	·51	·6950	·91	·8186	·31	·9049	·71	·9564	·11	·98257
·12	·5478	·52	·6985	·92	·8212	·32	·9066	·72	·9573	·12	·98300
·13	·5517	·53	·7019	·93	·8238	·33	·9082	·73	·9582	·13	·98341
·14	·5557	·54	·7054	·94	·8264	·34	·9099	·74	·9591	·14	·98382
0·15	0·5596	0·55	0·7088	0·95	0·8289	1·35	0·9115	1·75	0·9599	2·15	0·98422
·16	·5636	·56	·7123	·96	·8315	·36	·9131	·76	·9608	·16	·98461
·17	·5675	·57	·7157	·97	·8340	·37	·9147	·77	·9616	·17	·98500
·18	·5714	·58	·7190	·98	·8365	·38	·9162	·78	·9625	·18	·98537
·19	·5753	·59	·7224	·99	·8389	·39	·9177	·79	·9633	·19	·98574
0·20	0·5793	0·60	0·7257	1·00	0·8413	1·40	0·9192	1·80	0·9641	2·20	0·98610
·21	·5832	·61	·7291	·01	·8438	·41	·9207	·81	·9649	·21	·98645
·22	·5871	·62	·7324	·02	·8461	·42	·9222	·82	·9656	·22	·98679
·23	·5910	·63	·7357	·03	·8485	·43	·9236	·83	·9664	·23	·98713
·24	·5948	·64	·7389	·04	·8508	·44	·9251	·84	·9671	·24	·98745
0·25	0·5987	0·65	0·7422	1·05	0·8531	1·45	0·9265	1·85	0·9678	2·25	0·98778
·26	·6026	·66	·7454	·06	·8554	·46	·9279	·86	·9686	·26	·98809
·27	·6064	·67	·7486	·07	·8577	·47	·9292	·87	·9693	·27	·98840
·28	·6103	·68	·7517	·08	·8599	·48	·9306	·88	·9699	·28	·98870
·29	·6141	·69	·7549	·09	·8621	·49	·9319	·89	·9706	·29	·98899
0·30	0·6179	0·70	0·7580	1·10	0·8643	1·50	0·9332	1·90	0·9713	2·30	0·98928
·31	·6217	·71	·7611	·11	·8665	·51	·9345	·91	·9719	·31	·98956
·32	·6255	·72	·7642	·12	·8686	·52	·9357	·92	·9726	·32	·98983
·33	·6293	·73	·7673	·13	·8708	·53	·9370	·93	·9732	·33	·99010
·34	·6331	·74	·7704	·14	·8729	·54	·9382	·94	·9738	·34	·99036
0·35	0·6368	0·75	0·7734	1·15	0·8749	1·55	0·9394	1·95	0·9744	2·35	0·99061
·36	·6406	·76	·7764	·16	·8770	·56	·9406	·96	·9750	·36	·99086
·37	·6443	·77	·7794	·17	·8790	·57	·9418	·97	·9756	·37	·99111
·38	·6480	·78	·7823	·18	·8810	·58	·9429	·98	·9761	·38	·99134
·39	·6517	·79	·7852	·19	·8830	·59	·9441	·99	·9767	·39	·99158
0·40	0·6554	0·80	0·7881	1·20	0·8849	1·60	0·9452	2·00	0·9772	2·40	0·99180

TABLE 4. THE NORMAL DISTRIBUTION FUNCTION

x	$\Phi(x)$	x	$\Phi(x)$	x	$\Phi(x)$	x	$\Phi(x)$	x	$\Phi(x)$	x	$\Phi(x)$
2·40	0·99180	2·55	0·99461	2·70	0·99653	2·85	0·99781	3·00	0·99865	3·15	0·99918
·41	·99202	·56	·99477	·71	·99664	·86	·99788	·01	·99869	·16	·99921
·42	·99224	·57	·99492	·72	·99674	·87	·99795	·02	·99874	·17	·99924
·43	·99245	·58	·99506	·73	·99683	·88	·99801	·03	·99878	·18	·99926
·44	·99266	·59	·99520	·74	·99693	·89	·99807	·04	·99882	·19	·99929
2·45	0·99286	2·60	0·99534	2·75	0·99702	2·90	0·99813	3·05	0·99886	3·20	0·99931
·46	·99305	·61	·99547	·76	·99711	·91	·99819	·06	·99889	·21	·99934
·47	·99324	·62	·99560	·77	·99720	·92	·99825	·07	·99893	·22	·99936
·48	·99343	·63	·99573	·78	·99728	·93	·99831	·08	·99896	·23	·99938
·49	·99361	·64	·99585	·79	·99736	·94	·99836	·09	·99900	·24	·99940
2·50	0·99379	2·65	0·99598	2·80	0·99744	2·95	0·99841	3·10	0·99903	3·25	0·99942
·51	·99396	·66	·99609	·81	·99752	·96	·99846	·11	·99906	·26	·99944
·52	·99413	·67	·99621	·82	·99760	·97	·99851	·12	·99910	·27	·99946
·53	·99430	·68	·99632	·83	·99767	·98	·99856	·13	·99913	·28	·99948
·54	·99446	·69	·99643	·84	·99774	·99	·99861	·14	·99916	·29	·99950
2·55	0·99461	2·70	0·99653	2·85	0·99781	3·00	0·99865	3·15	0·99918	3·30	0·99952

The critical table below gives on the left the range of values of x for which $\Phi(x)$ takes the value on the right, correct to the last figure given; in critical cases, take the upper of the two values of $\Phi(x)$ indicated.

x	$\Phi(x)$	x	$\Phi(x)$	x	$\Phi(x)$	x	$\Phi(x)$
3·075	0·9990	3·263	0·9994	3·731	0·99990	3·916	0·99995
3·105	0·9991	3·320	0·9995	3·759	0·99991	3·976	0·99996
3·138	0·9992	3·389	0·9996	3·791	0·99992	4·055	0·99997
3·174	0·9993	3·480	0·9997	3·826	0·99993	4·173	0·99998
3·215	0·9994	3·615	0·9998	3·867	0·99994	4·417	0·99999
			0·9999		0·99995		1·00000

When $x > 3\cdot3$ the formula $1 - \Phi(x) \doteq \dfrac{e^{-\frac{1}{2}x^2}}{x\sqrt{2\pi}} \left[1 - \dfrac{1}{x^2} + \dfrac{3}{x^4} - \dfrac{15}{x^6} + \dfrac{105}{x^8} \right]$ is very accurate, with relative error less than $945/x^{10}$.

TABLE 5. PERCENTAGE POINTS OF THE NORMAL DISTRIBUTION

This table gives percentage points $x(P)$ defined by the equation

$$\frac{P}{100} = \frac{1}{\sqrt{2\pi}} \int_{x(P)}^{\infty} e^{-\frac{1}{2}t^2}\, dt.$$

If X is a variable, normally distributed with zero mean and unit variance, $P/100$ is the probability that $X \geqslant x(P)$. The lower P per cent points are given by symmetry as $-x(P)$, and the probability that $|X| \geqslant x(P)$ is $2P/100$.

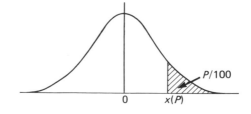

P	$x(P)$	P	$x(P)$	P	$x(P)$	P	$x(P)$	P	$x(P)$	P	$x(P)$
50	0·0000	5·0	1·6449	3·0	1·8808	2·0	2·0537	1·0	2·3263	0·10	3·0902
45	0·1257	4·8	1·6646	2·9	1·8957	1·9	2·0749	0·9	2·3656	0·09	3·1214
40	0·2533	4·6	1·6849	2·8	1·9110	1·8	2·0969	0·8	2·4089	0·08	3·1559
35	0·3853	4·4	1·7060	2·7	1·9268	1·7	2·1201	0·7	2·4573	0·07	3·1947
30	0·5244	4·2	1·7279	2·6	1·9431	1·6	2·1444	0·6	2·5121	0·06	3·2389
25	0·6745	4·0	1·7507	2·5	1·9600	1·5	2·1701	0·5	2·5758	0·05	3·2905
20	0·8416	3·8	1·7744	2·4	1·9774	1·4	2·1973	0·4	2·6521	0·01	3·7190
15	1·0364	3·6	1·7991	2·3	1·9954	1·3	2·2262	0·3	2·7478	0·005	3·8906
10	1·2816	3·4	1·8250	2·2	2·0141	1·2	2·2571	0·2	2·8782	0·001	4·2649
5	1·6449	3·2	1·8522	2·1	2·0335	1·1	2·2904	0·1	3·0902	0·0005	4·4172

TABLE 6. LOGARITHMS OF FACTORIALS

n	$\log_{10} n!$	n	$\log_{10} n!$	n	$\log_{10} n!$	n	$\log_{10} n!$	n	$\log_{10} n!$	n	$\log_{10} n!$
0	0·0000	50	64·4831	100	157·9700	150	262·7569	200	374·8969	250	492·5096
1	0·0000	51	66·1906	101	159·9743	151	264·9359	201	377·2001	251	494·9093
2	0·3010	52	67·9066	102	161·9829	152	267·1177	202	379·5054	252	497·3107
3	0·7782	53	69·6309	103	163·9958	153	269·3024	203	381·8129	253	499·7138
4	1·3802	54	71·3633	104	166·0128	154	271·4899	204	384·1226	254	502·1186
5	2·0792	55	73·1037	105	168·0340	155	273·6803	205	386·4343	255	504·5252
6	2·8573	56	74·8519	106	170·0593	156	275·8734	206	388·7482	256	506·9334
7	3·7024	57	76·6077	107	172·0887	157	278·0693	207	391·0642	257	509·3433
8	4·6055	58	78·3712	108	174·1221	158	280·2679	208	393·3822	258	511·7549
9	5·5598	59	80·1420	109	176·1595	159	282·4693	209	395·7024	259	514·1682
10	6·5598	60	81·9202	110	178·2009	160	284·6735	210	398·0246	260	516·5832
11	7·6012	61	83·7055	111	180·2462	161	286·8803	211	400·3489	261	518·9999
12	8·6803	62	85·4979	112	182·2955	162	289·0898	212	402·6752	262	521·4182
13	9·7943	63	87·2972	113	184·3485	163	291·3020	213	405·0036	263	523·8381
14	10·9404	64	89·1034	114	186·4054	164	293·5168	214	407·3340	264	526·2597
15	12·1165	65	90·9163	115	188·4661	165	295·7343	215	409·6664	265	528·6830
16	13·3206	66	92·7359	116	190·5306	166	297·9544	216	412·0009	266	531·1078
17	14·5511	67	94·5619	117	192·5988	167	300·1771	217	414·3373	267	533·5344
18	15·8063	68	96·3945	118	194·6707	168	302·4024	218	416·6758	268	535·9625
19	17·0851	69	98·2333	119	196·7462	169	304·6303	219	419·0162	269	538·3922
20	18·3861	70	100·0784	120	198·8254	170	306·8608	220	421·3587	270	540·8236
21	19·7083	71	101·9297	121	200·9082	171	309·0938	221	423·7031	271	543·2566
22	21·0508	72	103·7870	122	202·9945	172	311·3293	222	426·0494	272	545·6912
23	22·4125	73	105·6503	123	205·0844	173	313·5674	223	428·3977	273	548·1273
24	23·7927	74	107·5196	124	207·1779	174	315·8079	224	430·7480	274	550·5651
25	25·1906	75	109·3946	125	209·2748	175	318·0509	225	433·1002	275	553·0044
26	26·6056	76	111·2754	126	211·3751	176	320·2965	226	435·4543	276	555·4453
27	28·0370	77	113·1619	127	213·4790	177	322·5444	227	437·8103	277	557·8878
28	29·4841	78	115·0540	128	215·5862	178	324·7948	228	440·1682	278	560·3318
29	30·9465	79	116·9516	129	217·6967	179	327·0477	229	442·5281	279	562·7774
30	32·4237	80	118·8547	130	219·8107	180	329·3030	230	444·8898	280	565·2246
31	33·9150	81	120·7632	131	221·9280	181	331·5607	231	447·2534	281	567·6733
32	35·4202	82	122·6770	132	224·0485	182	333·8207	232	449·6189	282	570·1235
33	36·9387	83	124·5961	133	226·1724	183	336·0832	233	451·9862	283	572·5753
34	38·4702	84	126·5204	134	228·2995	184	338·3480	234	454·3555	284	575·0287
35	40·0142	85	128·4498	135	230·4298	185	340·6152	235	456·7265	285	577·4835
36	41·5705	86	130·3843	136	232·5634	186	342·8847	236	459·0994	286	579·9399
37	43·1387	87	132·3238	137	234·7001	187	345·1565	237	461·4742	287	582·3977
38	44·7185	88	134·2683	138	236·8400	188	347·4307	238	463·8508	288	584·8571
39	46·3096	89	136·2177	139	238·9830	189	349·7071	239	466·2292	289	587·3180
40	47·9116	90	138·1719	140	241·1291	190	351·9859	240	468·6094	290	589·7804
41	49·5244	91	140·1310	141	243·2783	191	354·2669	241	470·9914	291	592·2443
42	51·1477	92	142·0948	142	245·4306	192	356·5502	242	473·3752	292	594·7097
43	52·7811	93	144·0632	143	247·5860	193	358·8358	243	475·7608	293	597·1766
44	54·4246	94	146·0364	144	249·7443	194	361·1236	244	478·1482	294	599·6449
45	56·0778	95	148·0141	145	251·9057	195	363·4136	245	480·5374	295	602·1147
46	57·7406	96	149·9964	146	254·0700	196	365·7059	246	482·9283	296	604·5860
47	59·4127	97	151·9831	147	256·2374	197	368·0003	247	485·3210	297	607·0588
48	61·0939	98	153·9744	148	258·4076	198	370·2970	248	487·7154	298	609·5330
49	62·7841	99	155·9700	149	260·5808	199	372·5959	249	490·1116	299	612·0087
50	64·4831	100	157·9700	150	262·7569	200	374·8969	250	492·5096	300	614·4858

For large n, $\log_{10} n! \doteq 0·39909 + (n + \tfrac{1}{2}) \log_{10} n - 0·4342945\, n$.

TABLE 7. THE χ²-DISTRIBUTION FUNCTION

The function tabulated is

$$F_\nu(x) = \frac{1}{2^{\nu/2}\,\Gamma(\frac{\nu}{2})} \int_0^x t^{\frac12\nu-1}\, e^{-\frac12 t}\, dt$$

for integer $\nu \leqslant 25$. $F_\nu(x)$ is the probability that a random variable X, distributed as χ^2 with ν degrees of freedom, will be less than or equal to x. Note that $F_1(x) = 2\Phi(x^{\frac12}) - 1$ (cf. Table 4). For certain values of x and $\nu > 25$ use may be made of the following relation between the χ^2- and Poisson distributions:

$$F_\nu(x) = 1 - F(\tfrac12\nu - 1 \,|\, \tfrac12 x)$$

where $F(r|\mu)$ is the Poisson distribution function (see Table 2). If $\nu > 25$, X is approximately normally distributed

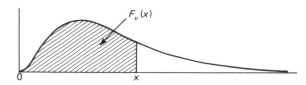

(The above shape applies for $\nu \geqslant 3$ only. When $\nu < 3$ the mode is at the origin.)

with mean ν and variance 2ν. A better approximation is usually obtained by using the formula

$$F_\nu(x) \doteqdot \Phi(\sqrt{2x} - \sqrt{2\nu - 1})$$

where $\Phi(s)$ is the normal distribution function (see Table 4).

Omitted entries to the left and right of tabulated values are 1 and 0 respectively (to four decimal places).

$\nu = 1$

x	F	x	F
0·0	0·0000	4·0	0·9545
·1	·2482	·1	·9571
·2	·3453	·2	·9596
·3	·4161	·3	·9619
·4	·4729	·4	·9641
0·5	0·5205	4·5	0·9661
·6	·5614	·6	·9680
·7	·5972	·7	·9698
·8	·6289	·8	·9715
·9	·6572	·9	·9731
1·0	0·6827	5·0	0·9747
·1	·7057	·1	·9761
·2	·7267	·2	·9774
·3	·7458	·3	·9787
·4	·7633	·4	·9799
1·5	0·7793	5·5	0·9810
·6	·7941	·6	·9820
·7	·8077	·7	·9830
·8	·8203	·8	·9840
·9	·8319	·9	·9849
2·0	0·8427	6·0	0·9857
·1	·8527	·1	·9865
·2	·8620	·2	·9872
·3	·8706	·3	·9879
·4	·8787	·4	·9886
2·5	0·8862	6·5	0·9892
·6	·8931	·6	·9898
·7	·8997	·7	·9904
·8	·9057	·8	·9909
·9	·9114	·9	·9914
3·0	0·9167	7·0	0·9918
·1	·9217	·1	·9923
·2	·9264	·2	·9927
·3	·9307	·3	·9931
·4	·9348	·4	·9935
3·5	0·9386	7·5	0·9938
·6	·9422	·6	·9942
·7	·9456	·7	·9945
·8	·9487	·8	·9948
·9	·9517	·9	·9951
4·0	0·9545	8·0	0·9953

$\nu = 2$

x	F	x	F
0·0	0·0000	4·0	0·8647
·1	·0488	·1	·8713
·2	·0952	·2	·8775
·3	·1393	·3	·8835
·4	·1813	·4	·8892
0·5	0·2212	4·5	0·8946
·6	·2592	·6	·8997
·7	·2953	·7	·9046
·8	·3297	·8	·9093
·9	·3624	·9	·9137
1·0	0·3935	5·0	0·9179
·1	·4231	·1	·9219
·2	·4512	·2	·9257
·3	·4780	·3	·9293
·4	·5034	·4	·9328
1·5	0·5276	5·5	0·9361
·6	·5507	·6	·9392
·7	·5726	·7	·9422
·8	·5934	·8	·9450
·9	·6133	·9	·9477
2·0	0·6321	6·0	0·9502
·1	·6501	·2	·9550
·2	·6671	·4	·9592
·3	·6834	·6	·9631
·4	·6988	·8	·9666
2·5	0·7135	7·0	0·9698
·6	·7275	·2	·9727
·7	·7408	·4	·9753
·8	·7534	·6	·9776
·9	·7654	·8	·9798
3·0	0·7769	8·0	0·9817
·1	·7878	·2	·9834
·2	·7981	·4	·9850
·3	·8080	·6	·9864
·4	·8173	·8	·9877
3·5	0·8262	9·0	0·9889
·6	·8347	·2	·9899
·7	·8428	·4	·9909
·8	·8504	·6	·9918
·9	·8577	·8	·9926
4·0	0·8647	10·0	0·9933

$\nu = 3$

x	F	x	F
0·0	0·0000	4·0	0·7385
·1	·0082	·2	·7593
·2	·0224	·4	·7786
·3	·0400	·6	·7965
·4	·0598	·8	·8130
0·5	0·0811	5·0	0·8282
·6	·1036	·2	·8423
·7	·1268	·4	·8553
·8	·1505	·6	·8672
·9	·1746	·8	·8782
1·0	0·1987	6·0	0·8884
·1	·2229	·2	·8977
·2	·2470	·4	·9063
·3	·2709	·6	·9142
·4	·2945	·8	·9214
1·5	0·3177	7·0	0·9281
·6	·3406	·2	·9342
·7	·3631	·4	·9398
·8	·3851	·6	·9450
·9	·4066	·8	·9497
2·0	0·4276	8·0	0·9540
·1	·4481	·2	·9579
·2	·4681	·4	·9616
·3	·4875	·6	·9649
·4	·5064	·8	·9679
2·5	0·5247	9·0	0·9707
·6	·5425	·2	·9733
·7	·5598	·4	·9756
·8	·5765	·6	·9777
·9	·5927	·8	·9797
3·0	0·6084	10·0	0·9814
·1	·6235	·2	·9831
·2	·6382	·4	·9845
·3	·6524	·6	·9859
·4	·6660	·8	·9871
3·5	0·6792	11·0	0·9883
·6	·6920	·2	·9893
·7	·7043	·4	·9903
·8	·7161	·6	·9911
·9	·7275	·8	·9919
4·0	0·7385	12·0	0·9926

TABLE 7. THE χ^2-DISTRIBUTION FUNCTION

$\nu =$	4	5	6	7	8	9	10	11	12	13	14
$x = 0.5$	0·0265	0·0079	0·0022	0·0006	0·0001						
1·0	·0902	·0374	·0144	·0052	·0018	0·0006	0·0002	0·0001			
1·5	·1734	·0869	·0405	·0177	·0073	·0029	·0011	·0004	0·0001		
2·0	·2642	·1509	·0803	·0402	·0190	·0085	·0037	·0015	·0006	0·0002	0·0001
2·5	0·3554	0·2235	0·1315	0·0729	0·0383	0·0191	0·0091	0·0042	0·0018	0·0008	0·0003
3·0	·4422	·3000	·1912	·1150	·0656	·0357	·0186	·0093	·0045	·0021	·0009
3·5	·5221	·3766	·2560	·1648	·1008	·0589	·0329	·0177	·0091	·0046	·0022
4·0	·5940	·4506	·3233	·2202	·1429	·0886	·0527	·0301	·0166	·0088	·0045
4·5	·6575	·5201	·3907	·2793	·1906	·1245	·0780	·0471	·0274	·0154	·0084
5·0	0·7127	0·5841	0·4562	0·3400	0·2424	0·1657	0·1088	0·0688	0·0420	0·0248	0·0142
5·5	·7603	·6421	·5185	·4008	·2970	·2113	·1446	·0954	·0608	·0375	·0224
6·0	·8009	·6938	·5768	·4603	·3528	·2601	·1847	·1266	·0839	·0538	·0335
6·5	·8352	·7394	·6304	·5173	·4086	·3110	·2283	·1620	·1112	·0739	·0477
7·0	·8641	·7794	·6792	·5711	·4634	·3629	·2746	·2009	·1424	·0978	·0653
7·5	0·8883	0·8140	0·7229	0·6213	0·5162	0·4148	0·3225	0·2427	0·1771	0·1254	0·0863
8·0	·9084	·8438	·7619	·6674	·5665	·4659	·3712	·2867	·2149	·1564	·1107
8·5	·9251	·8693	·7963	·7094	·6138	·5154	·4199	·3321	·2551	·1904	·1383
9·0	·9389	·8909	·8264	·7473	·6577	·5627	·4679	·3781	·2971	·2271	·1689
9·5	·9503	·9093	·8527	·7813	·6981	·6075	·5146	·4242	·3403	·2658	·2022
10·0	0·9596	0·9248	0·8753	0·8114	0·7350	0·6495	0·5595	0·4696	0·3840	0·3061	0·2378
10·5	·9672	·9378	·8949	·8380	·7683	·6885	·6022	·5140	·4278	·3474	·2752
11·0	·9734	·9486	·9116	·8614	·7983	·7243	·6425	·5567	·4711	·3892	·3140
11·5	·9785	·9577	·9259	·8818	·8251	·7570	·6801	·5976	·5134	·4310	·3536
12·0	·9826	·9652	·9380	·8994	·8488	·7867	·7149	·6364	·5543	·4724	·3937
12·5	0·9860	0·9715	0·9483	0·9147	0·8697	0·8134	0·7470	0·6727	0·5936	0·5129	0·4338
13·0	·9887	·9766	·9570	·9279	·8882	·8374	·7763	·7067	·6310	·5522	·4735
13·5	·9909	·9809	·9643	·9392	·9042	·8587	·8030	·7381	·6662	·5900	·5124
14·0	·9927	·9844	·9704	·9488	·9182	·8777	·8270	·7670	·6993	·6262	·5503
14·5	·9941	·9873	·9755	·9570	·9304	·8944	·8486	·7935	·7301	·6604	·5868
15·0	0·9953	0·9896	0·9797	0·9640	0·9409	0·9091	0·8679	0·8175	0·7586	0·6926	0·6218
15·5	·9962	·9916	·9833	·9699	·9499	·9219	·8851	·8393	·7848	·7228	·6551
16·0	·9970	·9932	·9862	·9749	·9576	·9331	·9004	·8589	·8088	·7509	·6866
16·5	·9976	·9944	·9887	·9791	·9642	·9429	·9138	·8764	·8306	·7768	·7162
17·0	·9981	·9955	·9907	·9826	·9699	·9513	·9256	·8921	·8504	·8007	·7438
17·5	0·9985	0·9964	0·9924	0·9856	0·9747	0·9586	0·9360	0·9061	0·8683	0·8226	0·7695
18·0	·9988	·9971	·9938	·9880	·9788	·9648	·9450	·9184	·8843	·8425	·7932
18·5	·9990	·9976	·9949	·9901	·9822	·9702	·9529	·9293	·8987	·8606	·8151
19·0	·9992	·9981	·9958	·9918	·9851	·9748	·9597	·9389	·9115	·8769	·8351
19·5	·9994	·9984	·9966	·9932	·9876	·9787	·9656	·9473	·9228	·8916	·8533
20	0·9995	0·9988	0·9972	0·9944	0·9897	0·9821	0·9707	0·9547	0·9329	0·9048	0·8699
21	·9997	·9992	·9982	·9962	·9929	·9873	·9789	·9666	·9496	·9271	·8984
22	·9998	·9995	·9988	·9975	·9951	·9911	·9849	·9756	·9625	·9446	·9214
23	·9999	·9997	·9992	·9983	·9966	·9938	·9893	·9823	·9723	·9583	·9397
24	·9999	·9998	·9995	·9989	·9977	·9957	·9924	·9873	·9797	·9689	·9542
25	0·9999	0·9999	0·9997	0·9992	0·9984	0·9970	0·9947	0·9909	0·9852	0·9769	0·9654
26		·9999	·9998	·9995	·9989	·9980	·9963	·9935	·9893	·9830	·9741
27		·9999	·9999	·9997	·9993	·9986	·9974	·9954	·9923	·9876	·9807
28			·9999	·9998	·9995	·9990	·9982	·9968	·9945	·9910	·9858
29			·9999	·9999	·9997	·9994	·9988	·9977	·9961	·9935	·9895
30				0·9999	0·9998	0·9996	0·9991	0·9984	0·9972	0·9953	0·9924

TABLE 7. THE χ^2-DISTRIBUTION FUNCTION

$\nu =$	15	16	17	18	19	20	21	22	23	24	25
$z = 3$	0.0004	0.0002	0.0001								
4	.0023	.0011	.0005	0.0002	0.0001						
5	0.0079	0.0042	0.0022	0.0011	0.0006	0.0003	0.0001	0.0001			
6	.0203	.0119	.0068	.0038	.0021	.0011	.0006	.0003	0.0001	0.0001	
7	.0424	.0267	.0165	.0099	.0058	.0033	.0019	.0010	.0005	.0003	0.0001
8	.0762	.0511	.0335	.0214	.0133	.0081	.0049	.0028	.0016	.0009	.0005
9	.1225	.0866	.0597	.0403	.0265	.0171	.0108	.0067	.0040	.0024	.0014
10	0.1803	0.1334	0.0964	0.0681	0.0471	0.0318	0.0211	0.0137	0.0087	0.0055	0.0033
11	.2474	.1905	.1434	.1056	.0762	.0538	.0372	.0253	.0168	.0110	.0071
12	.3210	.2560	.1999	.1528	.1144	.0839	.0604	.0426	.0295	.0201	.0134
13	.3977	.3272	.2638	.2084	.1614	.1226	.0914	.0668	.0480	.0339	.0235
14	.4745	.4013	.3329	.2709	.2163	.1695	.1304	.0985	.0731	.0533	.0383
15	0.5486	0.4754	0.4045	0.3380	0.2774	0.2236	0.1770	0.1378	0.1054	0.0792	0.0586
16	.6179	.5470	.4762	.4075	.3427	.2834	.2303	.1841	.1447	.1119	.0852
17	.6811	.6144	.5456	.4769	.4101	.3470	.2889	.2366	.1907	.1513	.1182
18	.7373	.6761	.6112	.5443	.4776	.4126	.3510	.2940	.2425	.1970	.1576
19	.7863	.7313	.6715	.6082	.5432	.4782	.4149	.3547	.2988	.2480	.2029
20	0.8281	0.7798	0.7258	0.6672	0.6054	0.5421	0.4787	0.4170	0.3581	0.3032	0.2532
21	.8632	.8215	.7737	.7206	.6632	.6029	.5411	.4793	.4189	.3613	.3074
22	.8922	.8568	.8153	.7680	.7157	.6595	.6005	.5401	.4797	.4207	.3643
23	.9159	.8863	.8507	.8094	.7627	.7112	.6560	.5983	.5392	.4802	.4224
24	.9349	.9105	.8806	.8450	.8038	.7576	.7069	.6528	.5962	.5384	.4806
25	0.9501	0.9302	0.9053	0.8751	0.8395	0.7986	0.7528	0.7029	0.6497	0.5942	0.5376
26	.9620	.9460	.9255	.9002	.8698	.8342	.7936	.7483	.6991	.6468	.5924
27	.9713	.9585	.9419	.9210	.8953	.8647	.8291	.7888	.7440	.6955	.6441
28	.9784	.9684	.9551	.9379	.9166	.8906	.8598	.8243	.7842	.7400	.6921
29	.9839	.9761	.9655	.9516	.9340	.9122	.8860	.8551	.8197	.7799	.7361
30	0.9881	0.9820	0.9737	0.9626	0.9482	0.9301	0.9080	0.8815	0.8506	0.8152	0.7757
31	.9912	.9865	.9800	.9712	.9596	.9448	.9263	.9039	.8772	.8462	.8110
32	.9936	.9900	.9850	.9780	.9687	.9567	.9414	.9226	.8999	.8730	.8420
33	.9953	.9926	.9887	.9833	.9760	.9663	.9538	.9381	.9189	.8959	.8689
34	.9966	.9946	.9916	.9874	.9816	.9739	.9638	.9509	.9348	.9153	.8921
35	0.9975	0.9960	0.9938	0.9905	0.9860	0.9799	0.9718	0.9613	0.9480	0.9316	0.9118
36	.9982	.9971	.9954	.9929	.9894	.9846	.9781	.9696	.9587	.9451	.9284
37	.9987	.9979	.9966	.9948	.9921	.9883	.9832	.9763	.9675	.9562	.9423
38	.9991	.9985	.9975	.9961	.9941	.9911	.9871	.9817	.9745	.9653	.9537
39	.9994	.9989	.9982	.9972	.9956	.9933	.9902	.9859	.9802	.9727	.9632
40	0.9995	0.9992	0.9987	0.9979	0.9967	0.9950	0.9926	0.9892	0.9846	0.9786	0.9708
41	.9997	.9994	.9991	.9985	.9976	.9963	.9944	.9918	.9882	.9833	.9770
42	.9998	.9996	.9993	.9989	.9982	.9972	.9958	.9937	.9909	.9871	.9820
43	.9998	.9997	.9995	.9992	.9987	.9980	.9969	.9953	.9931	.9901	.9860
44	.9999	.9998	.9997	.9994	.9991	.9985	.9977	.9965	.9947	.9924	.9892
45	0.9999	0.9999	0.9998	0.9996	0.9993	0.9989	0.9983	0.9973	0.9960	0.9942	0.9916
46	.9999	.9999	.9998	.9997	.9995	.9992	.9987	.9980	.9970	.9956	.9936
47		.9999	.9999	.9998	.9996	.9994	.9991	.9985	.9978	.9967	.9951
48			.9999	.9998	.9997	.9996	.9993	.9989	.9983	.9975	.9963
49			.9999	.9999	.9998	.9997	.9995	.9992	.9988	.9981	.9972
50				0.9999	0.9999	0.9998	0.9996	0.9994	0.9991	0.9986	0.9979

TABLE 8. PERCENTAGE POINTS OF THE χ^2-DISTRIBUTION

This table gives percentage points $\chi^2_\nu(P)$ defined by the equation

$$\frac{P}{100} = \frac{1}{2^{\nu/2}\,\Gamma(\frac{\nu}{2})} \int_{\chi^2_\nu(P)}^{\infty} x^{\frac{1}{2}\nu-1}\, e^{-\frac{1}{2}x}\, dx.$$

If X is a variable distributed as χ^2 with ν degrees of freedom, $P/100$ is the probability that $X \geqslant \chi^2_\nu(P)$.

For $\nu > 100$, $\sqrt{2X}$ is approximately normally distributed with mean $\sqrt{2\nu-1}$ and unit variance.

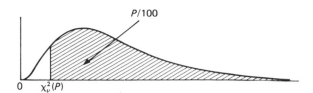

(The above shape applies for $\nu \geqslant 3$ only. When $\nu < 3$ the mode is at the origin.)

P	99·95	99·9	99·5	99	97·5	95	90	80	70	60
$\nu = 1$	0·0⁶3927	0·0⁵1571	0·0⁴3927	0·0³1571	0·0³9821	0·003932	0·01579	0·06418	0·1485	0·2750
2	0·001000	0·002001	0·01003	0·02010	0·05064	0·1026	0·2107	0·4463	0·7133	1·022
3	0·01528	0·02430	0·07172	0·1148	0·2158	0·3518	0·5844	1·005	1·424	1·869
4	0·06392	0·09080	0·2070	0·2971	0·4844	0·7107	1·064	1·649	2·195	2·753
5	0·1581	0·2102	0·4117	0·5543	0·8312	1·145	1·610	2·343	3·000	3·655
6	0·2994	0·3811	0·6757	0·8721	1·237	1·635	2·204	3·070	3·828	4·570
7	0·4849	0·5985	0·9893	1·239	1·690	2·167	2·833	3·822	4·671	5·493
8	0·7104	0·8571	1·344	1·646	2·180	2·733	3·490	4·594	5·527	6·423
9	0·9717	1·152	1·735	2·088	2·700	3·325	4·168	5·380	6·393	7·357
10	1·265	1·479	2·156	2·558	3·247	3·940	4·865	6·179	7·267	8·295
11	1·587	1·834	2·603	3·053	3·816	4·575	5·578	6·989	8·148	9·237
12	1·934	2·214	3·074	3·571	4·404	5·226	6·304	7·807	9·034	10·18
13	2·305	2·617	3·565	4·107	5·009	5·892	7·042	8·634	9·926	11·13
14	2·697	3·041	4·075	4·660	5·629	6·571	7·790	9·467	10·82	12·08
15	3·108	3·483	4·601	5·229	6·262	7·261	8·547	10·31	11·72	13·03
16	3·536	3·942	5·142	5·812	6·908	7·962	9·312	11·15	12·62	13·98
17	3·980	4·416	5·697	6·408	7·564	8·672	10·09	12·00	13·53	14·94
18	4·439	4·905	6·265	7·015	8·231	9·390	10·86	12·86	14·44	15·89
19	4·912	5·407	6·844	7·633	8·907	10·12	11·65	13·72	15·35	16·85
20	5·398	5·921	7·434	8·260	9·591	10·85	12·44	14·58	16·27	17·81
21	5·896	6·447	8·034	8·897	10·28	11·59	13·24	15·44	17·18	18·77
22	6·404	6·983	8·643	9·542	10·98	12·34	14·04	16·31	18·10	19·73
23	6·924	7·529	9·260	10·20	11·69	13·09	14·85	17·19	19·02	20·69
24	7·453	8·085	9·886	10·86	12·40	13·85	15·66	18·06	19·94	21·65
25	7·991	8·649	10·52	11·52	13·12	14·61	16·47	18·94	20·87	22·62
26	8·538	9·222	11·16	12·20	13·84	15·38	17·29	19·82	21·79	23·58
27	9·093	9·803	11·81	12·88	14·57	16·15	18·11	20·70	22·72	24·54
28	9·656	10·39	12·46	13·56	15·31	16·93	18·94	21·59	23·65	25·51
29	10·23	10·99	13·12	14·26	16·05	17·71	19·77	22·48	24·58	26·48
30	10·80	11·59	13·79	14·95	16·79	18·49	20·60	23·36	25·51	27·44
32	11·98	12·81	15·13	16·36	18·29	20·07	22·27	25·15	27·37	29·38
34	13·18	14·06	16·50	17·79	19·81	21·66	23·95	26·94	29·24	31·31
36	14·40	15·32	17·89	19·23	21·34	23·27	25·64	28·73	31·12	33·25
38	15·64	16·61	19·29	20·69	22·88	24·88	27·34	30·54	32·99	35·19
40	16·91	17·92	20·71	22·16	24·43	26·51	29·05	32·34	34·87	37·13
50	23·46	24·67	27·99	29·71	32·36	34·76	37·69	41·45	44·31	46·86
60	30·34	31·74	35·53	37·48	40·48	43·19	46·46	50·64	53·81	56·62
70	37·47	39·04	43·28	45·44	48·76	51·74	55·33	59·90	63·35	66·40
80	44·79	46·52	51·17	53·54	57·15	60·39	64·28	69·21	72·92	76·19
90	52·28	54·16	59·20	61·75	65·65	69·13	73·29	78·56	82·51	85·99
100	59·90	61·92	67·33	70·06	74·22	77·93	82·36	87·95	92·13	95·81

TABLE 8. PERCENTAGE POINTS OF THE χ²-DISTRIBUTION

This table gives percentage points $\chi^2_\nu(P)$ defined by the equation

$$\frac{P}{100} = \frac{1}{2^{\nu/2}\,\Gamma(\frac{\nu}{2})} \int_{\chi^2_\nu(P)}^{\infty} x^{\frac{1}{2}\nu-1}\, e^{-\frac{1}{2}x}\, dx.$$

If X is a variable distributed as χ^2 with ν degrees of freedom, $P/100$ is the probability that $X \geqslant \chi^2_\nu(P)$.

For $\nu > 100$, $\sqrt{2X}$ is approximately normally distributed with mean $\sqrt{2\nu-1}$ and unit variance.

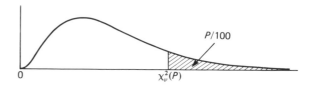

(The above shape applies for $\nu \geqslant 3$ only. When $\nu < 3$ the mode is at the origin.)

P	50	40	30	20	10	5	2·5	1	0·5	0·1	0·05
ν = 1	0·4549	0·7083	1·074	1·642	2·706	3·841	5·024	6·635	7·879	10·83	12·12
2	1·386	1·833	2·408	3·219	4·605	5·991	7·378	9·210	10·60	13·82	15·20
3	2·366	2·946	3·665	4·642	6·251	7·815	9·348	11·34	12·84	16·27	17·73
4	3·357	4·045	4·878	5·989	7·779	9·488	11·14	13·28	14·86	18·47	20·00
5	4·351	5·132	6·064	7·289	9·236	11·07	12·83	15·09	16·75	20·52	22·11
6	5·348	6·211	7·231	8·558	10·64	12·59	14·45	16·81	18·55	22·46	24·10
7	6·346	7·283	8·383	9·803	12·02	14·07	16·01	18·48	20·28	24·32	26·02
8	7·344	8·351	9·524	11·03	13·36	15·51	17·53	20·09	21·95	26·12	27·87
9	8·343	9·414	10·66	12·24	14·68	16·92	19·02	21·67	23·59	27·88	29·67
10	9·342	10·47	11·78	13·44	15·99	18·31	20·48	23·21	25·19	29·59	31·42
11	10·34	11·53	12·90	14·63	17·28	19·68	21·92	24·72	26·76	31·26	33·14
12	11·34	12·58	14·01	15·81	18·55	21·03	23·34	26·22	28·30	32·91	34·82
13	12·34	13·64	15·12	16·98	19·81	22·36	24·74	27·69	29·82	34·53	36·48
14	13·34	14·69	16·22	18·15	21·06	23·68	26·12	29·14	31·32	36·12	38·11
15	14·34	15·73	17·32	19·31	22·31	25·00	27·49	30·58	32·80	37·70	39·72
16	15·34	16·78	18·42	20·47	23·54	26·30	28·85	32·00	34·27	39·25	41·31
17	16·34	17·82	19·51	21·61	24·77	27·59	30·19	33·41	35·72	40·79	42·88
18	17·34	18·87	20·60	22·76	25·99	28·87	31·53	34·81	37·16	42·31	44·43
19	18·34	19·91	21·69	23·90	27·20	30·14	32·85	36·19	38·58	43·82	45·97
20	19·34	20·95	22·77	25·04	28·41	31·41	34·17	37·57	40·00	45·31	47·50
21	20·34	21·99	23·86	26·17	29·62	32·67	35·48	38·93	41·40	46·80	49·01
22	21·34	23·03	24·94	27·30	30·81	33·92	36·78	40·29	42·80	48·27	50·51
23	22·34	24·07	26·02	28·43	32·01	35·17	38·08	41·64	44·18	49·73	52·00
24	23·34	25·11	27·10	29·55	33·20	36·42	39·36	42·98	45·56	51·18	53·48
25	24·34	26·14	28·17	30·68	34·38	37·65	40·65	44·31	46·93	52·62	54·95
26	25·34	27·18	29·25	31·79	35·56	38·89	41·92	45·64	48·29	54·05	56·41
27	26·34	28·21	30·32	32·91	36·74	40·11	43·19	46·96	49·64	55·48	57·86
28	27·34	29·25	31·39	34·03	37·92	41·34	44·46	48·28	50·99	56·89	59·30
29	28·34	30·28	32·46	35·14	39·09	42·56	45·72	49·59	52·34	58·30	60·73
30	29·34	31·32	33·53	36·25	40·26	43·77	46·98	50·89	53·67	59·70	62·16
32	31·34	33·38	35·66	38·47	42·58	46·19	49·48	53·49	56·33	62·49	65·00
34	33·34	35·44	37·80	40·68	44·90	48·60	51·97	56·06	58·96	65·25	67·80
36	35·34	37·50	39·92	42·88	47·21	51·00	54·44	58·62	61·58	67·99	70·59
38	37·34	39·56	42·05	45·08	49·51	53·38	56·90	61·16	64·18	70·70	73·35
40	39·34	41·62	44·16	47·27	51·81	55·76	59·34	63·69	66·77	73·40	76·09
50	49·33	51·89	54·72	58·16	63·17	67·50	71·42	76·15	79·49	86·66	89·56
60	59·33	62·13	65·23	68·97	74·40	79·08	83·30	88·38	91·95	99·61	102·7
70	69·33	72·36	75·69	79·71	85·53	90·53	95·02	100·4	104·2	112·3	115·6
80	79·33	82·57	86·12	90·41	96·58	101·9	106·6	112·3	116·3	124·8	128·3
90	89·33	92·76	96·52	101·1	107·6	113·1	118·1	124·1	128·3	137·2	140·8
100	99·33	102·9	106·9	111·7	118·5	124·3	129·6	135·8	140·2	149·4	153·2

TABLE 9. THE *t*-DISTRIBUTION FUNCTION

The function tabulated is

$$F_\nu(t) = \frac{1}{\sqrt{\nu\pi}}\,\frac{\Gamma(\tfrac12\nu+\tfrac12)}{\Gamma(\tfrac12\nu)}\int_{-\infty}^{t}\frac{ds}{(1+s^2/\nu)^{\frac12(\nu+1)}}.$$

$F_\nu(t)$ is the probability that a random variable, distributed as t with ν degrees of freedom, will be less than or equal to t. When $t < 0$ use $F_\nu(t) = 1 - F_\nu(-t)$, the t distribution being symmetric about zero.

The limiting distribution of t as ν tends to infinity is the normal distribution with zero mean and unit variance (see Table 4). When ν is large interpolation in ν should be harmonic.

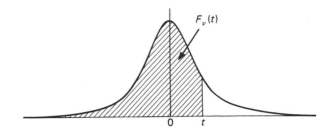

Omitted entries to the right of tabulated values are 1 (to four decimal places).

t	$\nu=1$	t	$\nu=1$	t	$\nu=2$	t	$\nu=2$	t	$\nu=3$	t	$\nu=3$
0·0	0·5000	4·0	0·9220	0·0	0·5000	4·0	0·9714	0·0	0·5000	4·0	0·9860
·1	·5317	4·2	·9256	·1	·5353	·1	·9727	·1	·5367	·1	·9869
·2	·5628	4·4	·9289	·2	·5700	·2	·9739	·2	·5729	·2	·9877
·3	·5928	4·6	·9319	·3	·6038	·3	·9750	·3	·6081	·3	·9884
·4	·6211	4·8	·9346	·4	·6361	·4	·9760	·4	·6420	·4	·9891
0·5	0·6476	5·0	0·9372	0·5	0·6667	4·5	0·9770	0·5	0·6743	4·5	0·9898
·6	·6720	5·5	·9428	·6	·6953	·6	·9779	·6	·7046	·6	·9903
·7	·6944	6·0	·9474	·7	·7218	·7	·9788	·7	·7328	·7	·9909
·8	·7148	6·5	·9514	·8	·7462	·8	·9796	·8	·7589	·8	·9914
·9	·7333	7·0	·9548	·9	·7684	·9	·9804	·9	·7828	·9	·9919
1·0	0·7500	7·5	0·9578	1·0	0·7887	5·0	0·9811	1·0	0·8045	5·0	0·9923
·1	·7651	8·0	·9604	·1	·8070	·1	·9818	·1	·8242	·1	·9927
·2	·7789	8·5	·9627	·2	·8235	·2	·9825	·2	·8419	·2	·9931
·3	·7913	9·0	·9648	·3	·8384	·3	·9831	·3	·8578	·3	·9934
·4	·8026	9·5	·9666	·4	·8518	·4	·9837	·4	·8720	·4	·9938
1·5	0·8128	10·0	0·9683	1·5	0·8638	5·5	0·9842	1·5	0·8847	5·5	0·9941
·6	·8222	10·5	·9698	·6	·8746	·6	·9848	·6	·8960	·6	·9944
·7	·8307	11·0	·9711	·7	·8844	·7	·9853	·7	·9062	·7	·9946
·8	·8386	11·5	·9724	·8	·8932	·8	·9858	·8	·9152	·8	·9949
·9	·8458	12·0	·9735	·9	·9011	·9	·9862	·9	·9232	·9	·9951
2·0	0·8524	12·5	0·9746	2·0	0·9082	6·0	0·9867	2·0	0·9303	6·0	0·9954
·1	·8585	13·0	·9756	·1	·9147	·1	·9871	·1	·9367	·1	·9956
·2	·8642	13·5	·9765	·2	·9206	·2	·9875	·2	·9424	·2	·9958
·3	·8695	14·0	·9773	·3	·9259	·3	·9879	·3	·9475	·3	·9960
·4	·8743	14·5	·9781	·4	·9308	·4	·9882	·4	·9521	·4	·9961
2·5	0·8789	15	0·9788	2·5	0·9352	6·5	0·9886	2·5	0·9561	6·5	0·9963
·6	·8831	16	·9801	·6	·9392	·6	·9889	·6	·9598	·6	·9965
·7	·8871	17	·9813	·7	·9429	·7	·9892	·7	·9631	·7	·9966
·8	·8908	18	·9823	·8	·9463	·8	·9895	·8	·9661	·8	·9967
·9	·8943	19	·9833	·9	·9494	·9	·9898	·9	·9687	·9	·9969
3·0	0·8976	20	0·9841	3·0	0·9523	7·0	0·9901	3·0	0·9712	7·0	0·9970
·1	·9007	21	·9849	·1	·9549	·1	·9904	·1	·9734	·1	·9971
·2	·9036	22	·9855	·2	·9573	·2	·9906	·2	·9753	·2	·9972
·3	·9063	23	·9862	·3	·9596	·3	·9909	·3	·9771	·3	·9973
·4	·9089	24	·9867	·4	·9617	·4	·9911	·4	·9788	·4	·9974
3·5	0·9114	25	0·9873	3·5	0·9636	7·5	0·9913	3·5	0·9803	7·5	0·9975
·6	·9138	30	·9894	·6	·9654	·6	·9916	·6	·9816	·6	·9976
·7	·9160	35	·9909	·7	·9670	·7	·9918	·7	·9829	·7	·9977
·8	·9181	40	·9920	·8	·9686	·8	·9920	·8	·9840	·8	·9978
·9	·9201	45	·9929	·9	·9701	·9	·9922	·9	·9850	·9	·9979
4·0	0·9220	50	0·9936	4·0	0·9714	8·0	0·9924	4·0	0·9860	8·0	0·9980

TABLE 9. THE *t*-DISTRIBUTION FUNCTION

$\nu =$	4	5	6	7	8	9	10	11	12	13	14
$t = 0.0$	0·5000	0·5000	0·5000	0·5000	0·5000	0·5000	0·5000	0·5000	0·5000	0·5000	0·5000
·1	·5374	·5379	·5382	·5384	·5386	·5387	·5388	·5389	·5390	·5391	·5391
·2	·5744	·5753	·5760	·5764	·5768	·5770	·5773	·5774	·5776	·5777	·5778
·3	·6104	·6119	·6129	·6136	·6141	·6145	·6148	·6151	·6153	·6155	·6157
·4	·6452	·6472	·6485	·6495	·6502	·6508	·6512	·6516	·6519	·6522	·6524
0·5	0·6783	0·6809	0·6826	0·6838	0·6847	0·6855	0·6861	0·6865	0·6869	0·6873	0·6876
·6	·7096	·7127	·7148	·7163	·7174	·7183	·7191	·7197	·7202	·7206	·7210
·7	·7387	·7424	·7449	·7467	·7481	·7492	·7501	·7508	·7514	·7519	·7523
·8	·7657	·7700	·7729	·7750	·7766	·7778	·7788	·7797	·7804	·7810	·7815
·9	·7905	·7953	·7986	·8010	·8028	·8042	·8054	·8063	·8071	·8078	·8083
1·0	0·8130	0·8184	0·8220	0·8247	0·8267	0·8283	0·8296	0·8306	0·8315	0·8322	0·8329
·1	·8335	·8393	·8433	·8461	·8483	·8501	·8514	·8526	·8535	·8544	·8551
·2	·8518	·8581	·8623	·8654	·8678	·8696	·8711	·8723	·8734	·8742	·8750
·3	·8683	·8748	·8793	·8826	·8851	·8870	·8886	·8899	·8910	·8919	·8927
·4	·8829	·8898	·8945	·8979	·9005	·9025	·9041	·9055	·9066	·9075	·9084
1·5	0·8960	0·9030	0·9079	0·9114	0·9140	0·9161	0·9177	0·9191	0·9203	0·9212	0·9221
·6	·9076	·9148	·9196	·9232	·9259	·9280	·9297	·9310	·9322	·9332	·9340
·7	·9178	·9251	·9300	·9335	·9362	·9383	·9400	·9414	·9426	·9435	·9444
·8	·9269	·9341	·9390	·9426	·9452	·9473	·9490	·9503	·9515	·9525	·9533
·9	·9349	·9421	·9469	·9504	·9530	·9551	·9567	·9580	·9591	·9601	·9609
2·0	0·9419	0·9490	0·9538	0·9572	0·9597	0·9617	0·9633	0·9646	0·9657	0·9666	0·9674
·1	·9482	·9551	·9598	·9631	·9655	·9674	·9690	·9702	·9712	·9721	·9728
·2	·9537	·9605	·9649	·9681	·9705	·9723	·9738	·9750	·9759	·9768	·9774
·3	·9585	·9651	·9694	·9725	·9748	·9765	·9779	·9790	·9799	·9807	·9813
·4	·9628	·9692	·9734	·9763	·9784	·9801	·9813	·9824	·9832	·9840	·9846
2·5	0·9666	0·9728	0·9767	0·9795	0·9815	0·9831	0·9843	0·9852	0·9860	0·9867	0·9873
·6	·9700	·9759	·9797	·9823	·9842	·9856	·9868	·9877	·9884	·9890	·9895
·7	·9730	·9786	·9822	·9847	·9865	·9878	·9888	·9897	·9903	·9909	·9914
·8	·9756	·9810	·9844	·9867	·9884	·9896	·9906	·9914	·9920	·9925	·9929
·9	·9779	·9831	·9863	·9885	·9901	·9912	·9921	·9928	·9933	·9938	·9942
3·0	0·9800	0·9850	0·9880	0·9900	0·9915	0·9925	0·9933	0·9940	0·9945	0·9949	0·9952
·1	·9819	·9866	·9894	·9913	·9927	·9936	·9944	·9949	·9954	·9958	·9961
·2	·9835	·9880	·9907	·9925	·9937	·9946	·9953	·9958	·9962	·9965	·9968
·3	·9850	·9893	·9918	·9934	·9946	·9954	·9960	·9965	·9968	·9971	·9974
·4	·9864	·9904	·9928	·9943	·9953	·9961	·9966	·9970	·9974	·9976	·9978
3·5	0·9876	0·9914	0·9936	0·9950	0·9960	0·9966	0·9971	0·9975	0·9978	0·9980	0·9982
·6	·9886	·9922	·9943	·9956	·9965	·9971	·9976	·9979	·9982	·9984	·9986
·7	·9896	·9930	·9950	·9962	·9970	·9975	·9979	·9982	·9985	·9987	·9988
·8	·9904	·9937	·9955	·9966	·9974	·9979	·9983	·9985	·9987	·9989	·9990
·9	·9912	·9943	·9960	·9971	·9977	·9982	·9985	·9988	·9989	·9991	·9992
4·0	0·9919	0·9948	0·9964	0·9974	0·9980	0·9984	0·9987	0·9990	0·9991	0·9992	0·9993
·1	·9926	·9953	·9968	·9977	·9983	·9987	·9989	·9991	·9993	·9994	·9995
·2	·9932	·9958	·9972	·9980	·9985	·9988	·9991	·9993	·9994	·9995	·9996
·3	·9937	·9961	·9975	·9982	·9987	·9990	·9992	·9994	·9995	·9996	·9996
·4	·9942	·9965	·9977	·9984	·9989	·9991	·9993	·9995	·9996	·9996	·9997
4·5	0·9946	0·9968	0·9979	0·9986	0·9990	0·9993	0·9994	0·9995	0·9996	0·9997	0·9998
·6	·9950	·9971	·9982	·9988	·9991	·9994	·9995	·9996	·9997	·9998	·9998
·7	·9953	·9973	·9983	·9989	·9992	·9994	·9996	·9997	·9997	·9998	·9998
·8	·9957	·9976	·9985	·9990	·9993	·9995	·9996	·9997	·9998	·9998	·9999
·9	·9960	·9978	·9986	·9991	·9994	·9996	·9997	·9998	·9998	·9999	·9999
5·0	0·9963	0·9979	0·9988	0·9992	0·9995	0·9996	0·9997	0·9998	0·9998	0·9999	0·9999

TABLE 9. THE t-DISTRIBUTION FUNCTION

$\nu =$	15	16	17	18	19	20	24	30	40	60	∞
$t = 0.0$	0.5000	0.5000	0.5000	0.5000	0.5000	0.5000	0.5000	0.5000	0.5000	0.5000	0.5000
.1	.5392	.5392	.5392	.5393	.5393	.5393	.5394	.5395	.5396	.5397	.5398
.2	.5779	.5780	.5781	.5781	.5782	.5782	.5784	.5786	.5788	.5789	.5793
.3	.6159	.6160	.6161	.6162	.6163	.6164	.6166	.6169	.6171	.6174	.6179
.4	.6526	.6528	.6529	.6531	.6532	.6533	.6537	.6540	.6544	.6547	.6554
0.5	0.6878	0.6881	0.6883	0.6884	0.6886	0.6887	0.6892	0.6896	0.6901	0.6905	0.6915
.6	.7213	.7215	.7218	.7220	.7222	.7224	.7229	.7235	.7241	.7246	.7257
.7	.7527	.7530	.7533	.7536	.7538	.7540	.7547	.7553	.7560	.7567	.7580
.8	.7819	.7823	.7826	.7829	.7832	.7834	.7842	.7850	.7858	.7866	.7881
.9	.8088	.8093	.8097	.8100	.8103	.8106	.8115	.8124	.8132	.8141	.8159
1.0	0.8334	0.8339	0.8343	0.8347	0.8351	0.8354	0.8364	0.8373	0.8383	0.8393	0.8413
.1	.8557	.8562	.8567	.8571	.8575	.8578	.8589	.8600	.8610	.8621	.8643
.2	.8756	.8762	.8767	.8772	.8776	.8779	.8791	.8802	.8814	.8826	.8849
.3	.8934	.8940	.8945	.8950	.8954	.8958	.8970	.8982	.8995	.9007	.9032
.4	.9091	.9097	.9103	.9107	.9112	.9116	.9128	.9141	.9154	.9167	.9192
1.5	0.9228	0.9235	0.9240	0.9245	0.9250	0.9254	0.9267	0.9280	0.9293	0.9306	0.9332
.6	.9348	.9354	.9360	.9365	.9370	.9374	.9387	.9400	.9413	.9426	.9452
.7	.9451	.9458	.9463	.9468	.9473	.9477	.9490	.9503	.9516	.9528	.9554
.8	.9540	.9546	.9552	.9557	.9561	.9565	.9578	.9590	.9603	.9616	.9641
.9	.9616	.9622	.9627	.9632	.9636	.9640	.9652	.9665	.9677	.9689	.9713
2.0	0.9680	0.9686	0.9691	0.9696	0.9700	0.9704	0.9715	0.9727	0.9738	0.9750	0.9772
.1	.9735	.9740	.9745	.9750	.9753	.9757	.9768	.9779	.9790	.9800	.9821
.2	.9781	.9786	.9790	.9794	.9798	.9801	.9812	.9822	.9832	.9842	.9861
.3	.9819	.9824	.9828	.9832	.9835	.9838	.9848	.9857	.9866	.9875	.9893
.4	.9851	.9855	.9859	.9863	.9866	.9869	.9877	.9886	.9894	.9902	.9918
2.5	0.9877	0.9882	0.9885	0.9888	0.9891	0.9894	0.9902	0.9909	0.9917	0.9924	0.9938
.6	.9900	.9903	.9907	.9910	.9912	.9914	.9921	.9928	.9935	.9941	.9953
.7	.9918	.9921	.9924	.9927	.9929	.9931	.9937	.9944	.9949	.9955	.9965
.8	.9933	.9936	.9938	.9941	.9943	.9945	.9950	.9956	.9961	.9966	.9974
.9	.9945	.9948	.9950	.9952	.9954	.9956	.9961	.9965	.9970	.9974	.9981
3.0	0.9955	0.9958	0.9960	0.9962	0.9963	0.9965	0.9969	0.9973	0.9977	0.9980	0.9987
.1	.9963	.9966	.9967	.9969	.9971	.9972	.9976	.9979	.9982	.9985	.9990
.2	.9970	.9972	.9974	.9975	.9976	.9978	.9981	.9984	.9987	.9989	.9993
.3	.9976	.9977	.9979	.9980	.9981	.9982	.9985	.9988	.9990	.9992	.9995
.4	.9980	.9982	.9983	.9984	.9985	.9986	.9988	.9990	.9992	.9994	.9997
3.5	0.9984	0.9985	0.9986	0.9987	0.9988	0.9989	0.9991	0.9993	0.9994	0.9996	0.9998
.6	.9987	.9988	.9989	.9990	.9990	.9991	.9993	.9994	.9996	.9997	.9998
.7	.9989	.9990	.9991	.9992	.9992	.9993	.9994	.9996	.9997	.9998	.9999
.8	.9991	.9992	.9993	.9993	.9994	.9994	.9996	.9997	.9998	.9998	.9999
.9	.9993	.9994	.9994	.9995	.9995	.9996	.9997	.9997	.9998	.9999	
4.0	0.9994	0.9995	0.9995	0.9996	0.9996	0.9996	0.9997	0.9998	0.9999	0.9999	
.1	.9995	.9996	.9996	.9997	.9997	.9997	.9998	.9999	.9999	.9999	
.2	.9996	.9997	.9997	.9997	.9998	.9998	.9998	.9999	.9999		
.3	.9997	.9997	.9998	.9998	.9998	.9998	.9999	.9999	.9999		
.4	.9997	.9998	.9998	.9998	.9998	.9999	.9999	.9999			
4.5	0.9998	0.9998	0.9998	0.9999	0.9999	0.9999	0.9999				

TABLE 10. PERCENTAGE POINTS OF THE *t*-DISTRIBUTION

This table gives percentage points $t_\nu(P)$ defined by the equation

$$\frac{P}{100} = \frac{1}{\sqrt{\nu\pi}} \frac{\Gamma(\frac{1}{2}\nu + \frac{1}{2})}{\Gamma(\frac{1}{2}\nu)} \int_{t_\nu(P)}^{\infty} \frac{dt}{(1 + t^2/\nu)^{\frac{1}{2}(\nu+1)}}.$$

Let X_1 and X_2 be independent random variables having a normal distribution with zero mean and unit variance and a χ^2-distribution with ν degrees of freedom respectively; then $t = X_1/\sqrt{X_2/\nu}$ has Student's t-distribution with ν degrees of freedom, and the probability that $t \geqslant t_\nu(P)$ is $P/100$. The lower percentage points are given by symmetry as $-t_\nu(P)$, and the probability that $|t| \geqslant t_\nu(P)$ is $2P/100$.

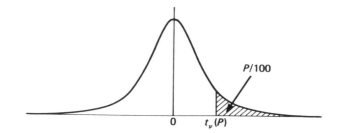

The limiting distribution of t as ν tends to infinity is the normal distribution with zero mean and unit variance. When ν is large interpolation in ν should be harmonic.

P	40	30	25	20	15	10	5	2·5	1	0·5	0·1	0·05
$\nu = 1$	0·3249	0·7265	1·0000	1·3764	1·963	3·078	6·314	12·71	31·82	63·66	318·3	636·6
2	0·2887	0·6172	0·8165	1·0607	1·386	1·886	2·920	4·303	6·965	9·925	22·33	31·60
3	0·2767	0·5844	0·7649	0·9785	1·250	1·638	2·353	3·182	4·541	5·841	10·21	12·92
4	0·2707	0·5686	0·7407	0·9410	1·190	1·533	2·132	2·776	3·747	4·604	7·173	8·610
5	0·2672	0·5594	0·7267	0·9195	1·156	1·476	2·015	2·571	3·365	4·032	5·893	6·869
6	0·2648	0·5534	0·7176	0·9057	1·134	1·440	1·943	2·447	3·143	3·707	5·208	5·959
7	0·2632	0·5491	0·7111	0·8960	1·119	1·415	1·895	2·365	2·998	3·499	4·785	5·408
8	0·2619	0·5459	0·7064	0·8889	1·108	1·397	1·860	2·306	2·896	3·355	4·501	5·041
9	0·2610	0·5435	0·7027	0·8834	1·100	1·383	1·833	2·262	2·821	3·250	4·297	4·781
10	0·2602	0·5415	0·6998	0·8791	1·093	1·372	1·812	2·228	2·764	3·169	4·144	4·587
11	0·2596	0·5399	0·6974	0·8755	1·088	1·363	1·796	2·201	2·718	3·106	4·025	4·437
12	0·2590	0·5386	0·6955	0·8726	1·083	1·356	1·782	2·179	2·681	3·055	3·930	4·318
13	0·2586	0·5375	0·6938	0·8702	1·079	1·350	1·771	2·160	2·650	3·012	3·852	4·221
14	0·2582	0·5366	0·6924	0·8681	1·076	1·345	1·761	2·145	2·624	2·977	3·787	4·140
15	0·2579	0·5357	0·6912	0·8662	1·074	1·341	1·753	2·131	2·602	2·947	3·733	4·073
16	0·2576	0·5350	0·6901	0·8647	1·071	1·337	1·746	2·120	2·583	2·921	3·686	4·015
17	0·2573	0·5344	0·6892	0·8633	1·069	1·333	1·740	2·110	2·567	2·898	3·646	3·965
18	0·2571	0·5338	0·6884	0·8620	1·067	1·330	1·734	2·101	2·552	2·878	3·610	3·922
19	0·2569	0·5333	0·6876	0·8610	1·066	1·328	1·729	2·093	2·539	2·861	3·579	3·883
20	0·2567	0·5329	0·6870	0·8600	1·064	1·325	1·725	2·086	2·528	2·845	3·552	3·850
21	0·2566	0·5325	0·6864	0·8591	1·063	1·323	1·721	2·080	2·518	2·831	3·527	3·819
22	0·2564	0·5321	0·6858	0·8583	1·061	1·321	1·717	2·074	2·508	2·819	3·505	3·792
23	0·2563	0·5317	0·6853	0·8575	1·060	1·319	1·714	2·069	2·500	2·807	3·485	3·768
24	0·2562	0·5314	0·6848	0·8569	1·059	1·318	1·711	2·064	2·492	2·797	3·467	3·745
25	0·2561	0·5312	0·6844	0·8562	1·058	1·316	1·708	2·060	2·485	2·787	3·450	3·725
26	0·2560	0·5309	0·6840	0·8557	1·058	1·315	1·706	2·056	2·479	2·779	3·435	3·707
27	0·2559	0·5306	0·6837	0·8551	1·057	1·314	1·703	2·052	2·473	2·771	3·421	3·690
28	0·2558	0·5304	0·6834	0·8546	1·056	1·313	1·701	2·048	2·467	2·763	3·408	3·674
29	0·2557	0·5302	0·6830	0·8542	1·055	1·311	1·699	2·045	2·462	2·756	3·396	3·659
30	0·2556	0·5300	0·6828	0·8538	1·055	1·310	1·697	2·042	2·457	2·750	3·385	3·646
32	0·2555	0·5297	0·6822	0·8530	1·054	1·309	1·694	2·037	2·449	2·738	3·365	3·622
34	0·2553	0·5294	0·6818	0·8523	1·052	1·307	1·691	2·032	2·441	2·728	3·348	3·601
36	0·2552	0·5291	0·6814	0·8517	1·052	1·306	1·688	2·028	2·434	2·719	3·333	3·582
38	0·2551	0·5288	0·6810	0·8512	1·051	1·304	1·686	2·024	2·429	2·712	3·319	3·566
40	0·2550	0·5286	0·6807	0·8507	1·050	1·303	1·684	2·021	2·423	2·704	3·307	3·551
50	0·2547	0·5278	0·6794	0·8489	1·047	1·299	1·676	2·009	2·403	2·678	3·261	3·496
60	0·2545	0·5272	0·6786	0·8477	1·045	1·296	1·671	2·000	2·390	2·660	3·232	3·460
120	0·2539	0·5258	0·6765	0·8446	1·041	1·289	1·658	1·980	2·358	2·617	3·160	3·373
∞	0·2533	0·5244	0·6745	0·8416	1·036	1·282	1·645	1·960	2·326	2·576	3·090	3·291

TABLE 11(a). 2·5 PER CENT POINTS OF BEHRENS' DISTRIBUTION

	θ	0°	15°	30°	45°	60°	75°	90°
$\nu_2 = 1$	$\nu_1 = 1$	12·71	15·56	17·36	17·97	17·36	15·56	12·71
	2	12·71	12·41	11·54	10·14	8·344	6·340	4·303
	3	12·71	12·29	11·11	9·303	7·123	4·960	3·182
	4	12·71	12·28	11·06	9·136	6·771	4·469	2·776
	5	12·71	12·28	11·04	9·090	6·636	4·218	2·571
	6	12·71	12·28	11·04	9·073	6·577	4·074	2·447
	7	12·71	12·28	11·04	9·065	6·546	3·980	2·365
	8	12·71	12·28	11·03	9·060	6·529	3·917	2·306
	10	12·71	12·28	11·03	9·055	6·511	3·835	2·228
	12	12·71	12·28	11·03	9·052	6·501	3·786	2·179
	24	12·71	12·28	11·03	9·046	6·485	3·685	2·064
	∞	12·71	12·28	11·03	9·040	6·473	3·615	1·960
$\nu_2 = 2$	$\nu_1 = 2$	4·303	4·414	4·563	4·624	4·563	4·414	4·303
	3	4·303	4·240	4·100	3·903	3·645	3·360	3·182
	4	4·303	4·205	3·964	3·653	3·312	2·978	2·776
	5	4·303	4·194	3·909	3·535	3·145	2·784	2·571
	6	4·303	4·190	3·882	3·468	3·045	2·667	2·447
	7	4·303	4·187	3·867	3·427	2·979	2·589	2·365
	8	4·303	4·186	3·857	3·400	2·933	2·534	2·306
	10	4·303	4·184	3·846	3·366	2·873	2·460	2·228
	12	4·303	4·182	3·840	3·346	2·835	2·414	2·179
	24	4·303	4·180	3·828	3·306	2·750	2·305	2·064
	∞	4·303	4·178	3·818	3·276	2·679	2·206	1·960
$\nu_2 = 3$	$\nu_1 = 3$	3·182	3·191	3·225	3·244	3·225	3·191	3·182
	4	3·182	3·149	3·088	3·012	2·913	2·816	2·776
	5	3·182	3·134	3·026	2·897	2·756	2·626	2·571
	6	3·182	3·127	2·992	2·831	2·663	2·513	2·447
	7	3·182	3·122	2·972	2·787	2·600	2·437	2·365
	8	3·182	3·120	2·958	2·758	2·556	2·384	2·306
	10	3·182	3·117	2·942	2·719	2·498	2·312	2·228
	12	3·182	3·115	2·933	2·696	2·462	2·267	2·179
	24	3·182	3·111	2·913	2·644	2·378	2·162	2·064
	∞	3·182	3·108	2·898	2·603	2·304	2·067	1·960
$\nu_2 = 4$	$\nu_1 = 4$	2·776	2·772	2·779	2·787	2·779	2·772	2·776
	5	2·776	2·754	2·717	2·675	2·625	2·582	2·571
	6	2·776	2·746	2·682	2·610	2·532	2·468	2·447
	7	2·776	2·741	2·660	2·567	2·471	2·392	2·365
	8	2·776	2·738	2·646	2·537	2·428	2·339	2·306
	10	2·776	2·734	2·628	2·498	2·371	2·268	2·228
	12	2·776	2·732	2·617	2·475	2·335	2·223	2·179
	24	2·776	2·727	2·594	2·421	2·252	2·118	2·064
	∞	2·776	2·723	2·576	2·377	2·178	2·024	1·960

If t_1 and t_2 are two independent random variables distributed as t with ν_1, ν_2 degrees of freedom respectively, the random variable $d = t_1 \sin \theta - t_2 \cos \theta$ has Behrens' distribution with parameters ν_1, ν_2 and θ. The function tabulated in Table 11 is $d_P = d_P(\nu_1, \nu_2, \theta)$ such that

$$\Pr(d > d_P) = P/100$$

for $P = 2·5$ and $0·5$ and a range of values of ν_1 and ν_2 with $\nu_1 \geqslant \nu_2$. When $\nu_1 < \nu_2$ use the result that

$$d_P(\nu_1, \nu_2, \theta) = d_P(\nu_2, \nu_1, 90° - \theta).$$

Behrens' distribution is symmetric about zero, so

$$\Pr(|d| > d_P) = 2P/100.$$

Notice that in this table θ is measured in degrees rather than radians.

TABLE 11(a). 2·5 PER CENT POINTS OF BEHRENS' DISTRIBUTION

	θ	0°	15°	30°	45°	60°	75°	90°
$\nu_2 = 5$	$\nu_1 = 5$	2·571	2·564	2·562	2·565	2·562	2·564	2·571
	6	2·571	2·554	2·527	2·500	2·470	2·449	2·447
	7	2·571	2·549	2·505	2·458	2·410	2·374	2·365
	8	2·571	2·546	2·490	2·428	2·367	2·320	2·306
	10	2·571	2·541	2·471	2·390	2·310	2·248	2·228
	12	2·571	2·539	2·460	2·366	2·274	2·203	2·179
	24	2·571	2·533	2·436	2·312	2·191	2·098	2·064
	∞	2·571	2·529	2·416	2·266	2·118	2·004	1·960
$\nu_2 = 6$	$\nu_1 = 6$	2·447	2·440	2·435	2·436	2·435	2·440	2·447
	7	2·447	2·434	2·413	2·394	2·375	2·364	2·365
	8	2·447	2·431	2·398	2·364	2·331	2·310	2·306
	10	2·447	2·426	2·379	2·325	2·274	2·238	2·228
	12	2·447	2·423	2·367	2·301	2·239	2·193	2·179
	24	2·447	2·418	2·342	2·247	2·156	2·088	2·064
	∞	2·447	2·413	2·322	2·201	2·082	1·993	1·960
$\nu_2 = 7$	$\nu_1 = 7$	2·365	2·358	2·352	2·352	2·352	2·358	2·365
	8	2·365	2·354	2·337	2·322	2·309	2·304	2·306
	10	2·365	2·350	2·317	2·283	2·252	2·232	2·228
	12	2·365	2·347	2·306	2·259	2·216	2·187	2·179
	24	2·365	2·341	2·280	2·205	2·133	2·082	2·064
	∞	2·365	2·336	2·259	2·158	2·060	1·987	1·960
$\nu_2 = 8$	$\nu_1 = 8$	2·306	2·300	2·294	2·292	2·294	2·300	2·306
	10	2·306	2·295	2·274	2·254	2·237	2·228	2·228
	12	2·306	2·292	2·262	2·230	2·201	2·183	2·179
	24	2·306	2·286	2·236	2·175	2·118	2·077	2·064
	∞	2·306	2·281	2·215	2·128	2·044	1·982	1·960
$\nu_2 = 10$	$\nu_1 = 10$	2·228	2·223	2·217	2·215	2·217	2·223	2·228
	12	2·228	2·220	2·205	2·191	2·181	2·178	2·179
	24	2·228	2·214	2·178	2·136	2·098	2·072	2·064
	∞	2·228	2·209	2·157	2·089	2·024	1·977	1·960
$\nu_2 = 12$	$\nu_1 = 12$	2·179	2·175	2·169	2·167	2·169	2·175	2·179
	24	2·179	2·168	2·142	2·112	2·085	2·069	2·064
	∞	2·179	2·163	2·120	2·064	2·011	1·973	1·960
$\nu_2 = 24$	$\nu_1 = 24$	2·064	2·062	2·058	2·056	2·058	2·062	2·064
	∞	2·064	2·056	2·035	2·009	1·983	1·966	1·960
$\nu_2 = \infty$	$\nu_1 = \infty$	1·960	1·960	1·960	1·960	1·960	1·960	1·960

This distribution arises in investigating the difference between the means μ_1, μ_2 of two normal distributions without assuming, as does the t-statistic, that the variances are equal. Let $\bar{x}_1$, $\bar{x}_2$ be the means and s_1^2, s_2^2 the variances of two independent samples of sizes n_1, n_2 from normal distributions, let $\nu_1 = n_1 - 1$, $\nu_2 = n_2 - 1$ and $\theta = \tan^{-1}\left(\dfrac{s_1}{\sqrt{\nu_1}} \middle/ \dfrac{s_2}{\sqrt{\nu_2}}\right)$, θ being measured in degrees. Define $r = \sqrt{\dfrac{s_1^2}{\nu_1} + \dfrac{s_2^2}{\nu_2}}$ and $d = \dfrac{\bar{x}_1 - \bar{x}_2}{r}$.

If $d > d_P$ the *confidence level* associated with $\mu_1 \leqslant \mu_2$ is less than P per cent, and if $d < -d_P$ the *confidence level* associated with $\mu_1 \geqslant \mu_2$ is less than P per cent. (See H. Cramér, *Mathematical Methods of Statistics*, Princeton University Press (1946), Princeton, N.J., pp. 520–523.) Also, the values of $\mu_1 - \mu_2$ such that $|(\bar{x}_1 - \bar{x}_2) - (\mu_1 - \mu_2)| \leqslant r d_P$ provide a $100 - 2P$ per cent *Bayesian credibility interval* for $\mu_1 - \mu_2$.

TABLE 11(b). 0·5 PER CENT POINTS OF BEHRENS' DISTRIBUTION

	θ	0°	15°	30°	45°	60°	75°	90°
$\nu_2 = 1$	$\nu_1 = 1$	63·66	77·96	86·96	90·02	86·96	77·96	63·66
	2	63·66	61·61	55·62	46·18	34·18	21·11	9·925
	3	63·66	61·49	55·15	45·08	32·04	17·28	5·841
	4	63·66	61·49	55·14	45·04	31·89	16·70	4·604
	5	63·66	61·49	55·14	45·03	31·87	16·59	4·032
	6	63·66	61·49	55·14	45·03	31·86	16·57	3·707
	7	63·66	61·49	55·13	45·03	31·86	16·56	3·499
	8	63·66	61·49	55·13	45·03	31·86	16·55	3·355
	10	63·66	61·49	55·13	45·03	31·86	16·55	3·169
	12	63·66	61·49	55·13	45·03	31·86	16·54	3·055
	24	63·66	61·49	55·13	45·02	31·85	16·54	2·797
	∞	63·66	61·49	55·13	45·02	31·85	16·53	2·576
$\nu_2 = 2$	$\nu_1 = 2$	9·925	10·01	10·14	10·19	10·14	10·01	9·925
	3	9·925	9·640	8·905	7·937	6·966	6·187	5·841
	4	9·925	9·609	8·717	7·428	6·082	5·049	4·604
	5	9·925	9·604	8·676	7·270	5·716	4·528	4·032
	6	9·925	9·602	8·663	7·210	5·535	4·235	3·707
	7	9·925	9·601	8·657	7·183	5·434	4·049	3·499
	8	9·925	9·600	8·653	7·169	5·373	3·921	3·355
	10	9·925	9·600	8·649	7·155	5·308	3·759	3·169
	12	9·925	9·599	8·647	7·148	5·275	3·660	3·055
	24	9·925	9·598	8·642	7·134	5·223	3·446	2·797
	∞	9·925	9·597	8·638	7·124	5·194	3·276	2·576
$\nu_2 = 3$	$\nu_1 = 3$	5·841	5·754	5·640	5·598	5·640	5·754	5·841
	4	5·841	5·694	5·349	4·986	4·720	4·601	4·604
	5	5·841	5·681	5·256	4·739	4·316	4·076	4·032
	6	5·841	5·676	5·218	4·617	4·095	3·782	3·707
	7	5·841	5·673	5·199	4·548	3·958	3·595	3·499
	8	5·841	5·671	5·189	4·506	3·866	3·467	3·355
	10	5·841	5·669	5·177	4·459	3·753	3·302	3·169
	12	5·841	5·668	5·171	4·434	3·686	3·201	3·055
	24	5·841	5·666	5·159	4·389	3·548	2·977	2·797
	∞	5·841	5·664	5·150	4·361	3·449	2·789	2·576
$\nu_2 = 4$	$\nu_1 = 4$	4·604	4·525	4·400	4·350	4·400	4·525	4·604
	5	4·604	4·505	4·283	4·084	3·983	3·993	4·032
	6	4·604	4·497	4·229	3·945	3·755	3·694	3·707
	7	4·604	4·493	4·201	3·862	3·613	3·504	3·499
	8	4·604	4·490	4·184	3·809	3·517	3·373	3·355
	10	4·604	4·487	4·165	3·745	3·395	3·206	3·169
	12	4·604	4·486	4·155	3·709	3·323	3·104	3·055
	24	4·604	4·482	4·135	3·640	3·167	2·876	2·797
	∞	4·604	4·479	4·121	3·592	3·045	2·685	2·576

If t_1 and t_2 are two independent random variables distributed as t with ν_1, ν_2 degrees of freedom respectively, the random variable $d = t_1 \sin\theta - t_2 \cos\theta$ has Behrens' distribution with parameters ν_1, ν_2 and θ. The function tabulated in Table 11 is $d_P = d_P(\nu_1, \nu_2, \theta)$ such that

$$\Pr(d > d_P) = P/100$$

for $P = 2\cdot5$ and $0\cdot5$ and a range of values of ν_1 and ν_2 with $\nu_1 \geqslant \nu_2$. When $\nu_1 < \nu_2$ use the result that

$$d_P(\nu_1, \nu_2, \theta) = d_P(\nu_2, \nu_1, 90° - \theta).$$

Behrens' distribution is symmetric about zero, so

$$\Pr(|d| > d_P) = 2P/100.$$

Notice that in this table θ is measured in degrees **rather than** radians.

TABLE 11(b). 0·5 PER CENT POINTS OF BEHRENS' DISTRIBUTION

		θ	0°	15°	30°	45°	60°	75°	90°
$\nu_2 = 5$	$\nu_1 = 5$		4·032	3·968	3·856	3·809	3·856	3·968	4·032
	6		4·032	3·957	3·794	3·663	3·622	3·666	3·707
	7		4·032	3·952	3·760	3·575	3·476	3·474	3·499
	8		4·032	3·949	3·739	3·518	3·378	3·342	3·355
	10		4·032	3·945	3·715	3·447	3·253	3·173	3·169
	12		4·032	3·943	3·702	3·407	3·178	3·069	3·055
	24		4·032	3·938	3·677	3·325	3·016	2·840	2·797
	∞		4·032	3·934	3·658	3·266	2·886	2·646	2·576
$\nu_2 = 6$	$\nu_1 = 6$		3·707	3·654	3·556	3·514	3·556	3·654	3·707
	7		3·707	3·648	3·519	3·423	3·408	3·461	3·499
	8		3·707	3·644	3·496	3·363	3·308	3·328	3·355
	10		3·707	3·639	3·468	3·289	3·180	3·158	3·169
	12		3·707	3·637	3·453	3·246	3·104	3·053	3·055
	24		3·707	3·631	3·424	3·158	2·938	2·822	2·797
	∞		3·707	3·627	3·402	3·093	2·804	2·627	2·576
$\nu_2 = 7$	$\nu_1 = 7$		3·499	3·454	3·369	3·331	3·369	3·454	3·499
	8		3·499	3·450	3·344	3·269	3·267	3·321	3·355
	10		3·499	3·445	3·314	3·193	3·138	3·149	3·169
	12		3·499	3·442	3·298	3·149	3·060	3·045	3·055
	24		3·499	3·436	3·265	3·056	2·892	2·812	2·797
	∞		3·499	3·431	3·241	2·987	2·755	2·616	2·576
$\nu_2 = 8$	$\nu_1 = 8$		3·355	3·316	3·241	3·206	3·241	3·316	3·355
	10		3·355	3·310	3·210	3·129	3·110	3·144	3·169
	12		3·355	3·307	3·192	3·083	3·032	3·039	3·055
	24		3·355	3·301	3·158	2·988	2·862	2·806	2·797
	∞		3·355	3·295	3·132	2·916	2·723	2·608	2·576
$\nu_2 = 10$	$\nu_1 = 10$		3·169	3·138	3·078	3·049	3·078	3·138	3·169
	12		3·169	3·135	3·059	3·002	2·998	3·033	3·055
	24		3·169	3·127	3·021	2·904	2·825	2·798	2·797
	∞		3·169	3·121	2·993	2·828	2·684	2·600	2·576
$\nu_2 = 12$	$\nu_1 = 12$		3·055	3·029	2·978	2·954	2·978	3·029	3·055
	24		3·055	3·021	2·939	2·853	2·803	2·794	2·797
	∞		3·055	3·015	2·909	2·775	2·661	2·595	2·576
$\nu_2 = 24$	$\nu_1 = 24$		2·797	2·784	2·759	2·747	2·759	2·784	2·797
	∞		2·797	2·777	2·726	2·664	2·613	2·584	2·576
$\nu_2 = \infty$	$\nu_1 = \infty$		2·576	2·576	2·576	2·576	2·576	2·576	2·576

This distribution arises in investigating the difference between the means μ_1, μ_2 of two normal distributions without assuming, as does the t-statistic, that the variances are equal. Let $\bar{x}_1$, $\bar{x}_2$ be the means and s_1^2, s_2^2 the variances of two independent samples of sizes n_1, n_2 from normal distributions, let $\nu_1 = n_1 - 1$, $\nu_2 = n_2 - 1$ and $\theta = \tan^{-1}\left(\frac{s_1}{\sqrt{\nu_1}}\bigg/\frac{s_2}{\sqrt{\nu_2}}\right)$, θ being measured in degrees. Define $r = \sqrt{\frac{s_1^2}{\nu_1} + \frac{s_2^2}{\nu_2}}$ and $d = \frac{\bar{x}_1 - \bar{x}_2}{r}$.

If $d > d_P$ the *confidence level* associated with $\mu_1 \leqslant \mu_2$ is less than P per cent, and if $d < -d_P$ the *confidence level* associated with $\mu_1 \geqslant \mu_2$ is less than P per cent. (See H. Cramér, *Mathematical Methods of Statistics*, Princeton University Press (1946), Princeton, N.J., pp. 520–523.) Also, the values of $\mu_1 - \mu_2$ such that $|(\bar{x}_1 - \bar{x}_2) - (\mu_1 - \mu_2)| \leqslant r d_P$ provide a $100 - 2P$ per cent *Bayesian credibility interval* for $\mu_1 - \mu_2$.

TABLE 12(*a*). 10 PER CENT POINTS OF THE *F*-DISTRIBUTION

The function tabulated is $F(P) = F(P|\nu_1, \nu_2)$ defined by the equation

$$\frac{P}{100} = \frac{\Gamma(\frac{1}{2}\nu_1 + \frac{1}{2}\nu_2)}{\Gamma(\frac{1}{2}\nu_1)\,\Gamma(\frac{1}{2}\nu_2)}\,\nu_1^{\frac{1}{2}\nu_1}\,\nu_2^{\frac{1}{2}\nu_2}\int_{F(P)}^{\infty}\frac{F^{\frac{1}{2}\nu_1-1}}{(\nu_2 + \nu_1 F)^{\frac{1}{2}(\nu_1+\nu_2)}}\,dF,$$

for $P = 10, 5, 2\cdot5, 1, 0\cdot5$ and $0\cdot1$. The lower percentage points, that is the values $F'(P) = F'(P|\nu_1, \nu_2)$ such that the probability that $F \leqslant F'(P)$ is equal to $P/100$, may be found by the formula

$$F'(P|\nu_1, \nu_2) = 1/F(P|\nu_2, \nu_1).$$

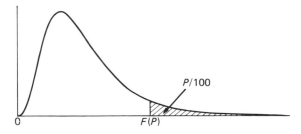

(This shape applies only when $\nu_1 \geqslant 3$. When $\nu_1 < 3$ the mode is at the origin.)

$\nu_1 =$	1	2	3	4	5	6	7	8	10	12	24	∞
$\nu_2 = 1$	39·86	49·50	53·59	55·83	57·24	58·20	58·91	59·44	60·19	60·71	62·00	63·33
2	8·526	9·000	9·162	9·243	9·293	9·326	9·349	9·367	9·392	9·408	9·450	9·491
3	5·538	5·462	5·391	5·343	5·309	5·285	5·266	5·252	5·230	5·216	5·176	5·134
4	4·545	4·325	4·191	4·107	4·051	4·010	3·979	3·955	3·920	3·896	3·831	3·761
5	4·060	3·780	3·619	3·520	3·453	3·405	3·368	3·339	3·297	3·268	3·191	3·105
6	3·776	3·463	3·289	3·181	3·108	3·055	3·014	2·983	2·937	2·905	2·818	2·722
7	3·589	3·257	3·074	2·961	2·883	2·827	2·785	2·752	2·703	2·668	2·575	2·471
8	3·458	3·113	2·924	2·806	2·726	2·668	2·624	2·589	2·538	2·502	2·404	2·293
9	3·360	3·006	2·813	2·693	2·611	2·551	2·505	2·469	2·416	2·379	2·277	2·159
10	3·285	2·924	2·728	2·605	2·522	2·461	2·414	2·377	2·323	2·284	2·178	2·055
11	3·225	2·860	2·660	2·536	2·451	2·389	2·342	2·304	2·248	2·209	2·100	1·972
12	3·177	2·807	2·606	2·480	2·394	2·331	2·283	2·245	2·188	2·147	2·036	1·904
13	3·136	2·763	2·560	2·434	2·347	2·283	2·234	2·195	2·138	2·097	1·983	1·846
14	3·102	2·726	2·522	2·395	2·307	2·243	2·193	2·154	2·095	2·054	1·938	1·797
15	3·073	2·695	2·490	2·361	2·273	2·208	2·158	2·119	2·059	2·017	1·899	1·755
16	3·048	2·668	2·462	2·333	2·244	2·178	2·128	2·088	2·028	1·985	1·866	1·718
17	3·026	2·645	2·437	2·308	2·218	2·152	2·102	2·061	2·001	1·958	1·836	1·686
18	3·007	2·624	2·416	2·286	2·196	2·130	2·079	2·038	1·977	1·933	1·810	1·657
19	2·990	2·606	2·397	2·266	2·176	2·109	2·058	2·017	1·956	1·912	1·787	1·631
20	2·975	2·589	2·380	2·249	2·158	2·091	2·040	1·999	1·937	1·892	1·767	1·607
21	2·961	2·575	2·365	2·233	2·142	2·075	2·023	1·982	1·920	1·875	1·748	1·586
22	2·949	2·561	2·351	2·219	2·128	2·060	2·008	1·967	1·904	1·859	1·731	1·567
23	2·937	2·549	2·339	2·207	2·115	2·047	1·995	1·953	1·890	1·845	1·716	1·549
24	2·927	2·538	2·327	2·195	2·103	2·035	1·983	1·941	1·877	1·832	1·702	1·533
25	2·918	2·528	2·317	2·184	2·092	2·024	1·971	1·929	1·866	1·820	1·689	1·518
26	2·909	2·519	2·307	2·174	2·082	2·014	1·961	1·919	1·855	1·809	1·677	1·504
27	2·901	2·511	2·299	2·165	2·073	2·005	1·952	1·909	1·845	1·799	1·666	1·491
28	2·894	2·503	2·291	2·157	2·064	1·996	1·943	1·900	1·836	1·790	1·656	1·478
29	2·887	2·495	2·283	2·149	2·057	1·988	1·935	1·892	1·827	1·781	1·647	1·467
30	2·881	2·489	2·276	2·142	2·049	1·980	1·927	1·884	1·819	1·773	1·638	1·456
32	2·869	2·477	2·263	2·129	2·036	1·967	1·913	1·870	1·805	1·758	1·622	1·437
34	2·859	2·466	2·252	2·118	2·024	1·955	1·901	1·858	1·793	1·745	1·608	1·419
36	2·850	2·456	2·243	2·108	2·014	1·945	1·891	1·847	1·781	1·734	1·595	1·404
38	2·842	2·448	2·234	2·099	2·005	1·935	1·881	1·838	1·772	1·724	1·584	1·390
40	2·835	2·440	2·226	2·091	1·997	1·927	1·873	1·829	1·763	1·715	1·574	1·377
60	2·791	2·393	2·177	2·041	1·946	1·875	1·819	1·775	1·707	1·657	1·511	1·291
120	2·748	2·347	2·130	1·992	1·896	1·824	1·767	1·722	1·652	1·601	1·447	1·193
∞	2·706	2·303	2·084	1·945	1·847	1·774	1·717	1·670	1·599	1·546	1·383	1·000

TABLE 12(b). 5 PER CENT POINTS OF THE F-DISTRIBUTION

If $F = \dfrac{X_1}{\nu_1} \Big/ \dfrac{X_2}{\nu_2}$, where X_1 and X_2 are independent random variables distributed as χ^2 with ν_1 and ν_2 degrees of freedom respectively, then the probabilities that $F \geqslant F(P)$ and that $F \leqslant F'(P)$ are both equal to $P/100$. Linear interpolation in ν_1 and ν_2 will generally be sufficiently accurate except when either $\nu_1 > 12$ or $\nu_2 > 40$, when harmonic interpolation should be used.

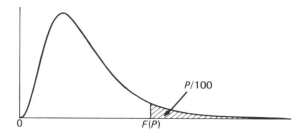

(This shape applies only when $\nu_1 \geqslant 3$. When $\nu_1 < 3$ the mode is at the origin.)

$\nu_1 =$	1	2	3	4	5	6	7	8	10	12	24	∞
$\nu_2 = 1$	161·4	199·5	215·7	224·6	230·2	234·0	236·8	238·9	241·9	243·9	249·1	254·3
2	18·51	19·00	19·16	19·25	19·30	19·33	19·35	19·37	19·40	19·41	19·45	19·50
3	10·13	9·552	9·277	9·117	9·013	8·941	8·887	8·845	8·786	8·745	8·639	8·526
4	7·709	6·944	6·591	6·388	6·256	6·163	6·094	6·041	5·964	5·912	5·774	5·628
5	6·608	5·786	5·409	5·192	5·050	4·950	4·876	4·818	4·735	4·678	4·527	4·365
6	5·987	5·143	4·757	4·534	4·387	4·284	4·207	4·147	4·060	4·000	3·841	3·669
7	5·591	4·737	4·347	4·120	3·972	3·866	3·787	3·726	3·637	3·575	3·410	3·230
8	5·318	4·459	4·066	3·838	3·687	3·581	3·500	3·438	3·347	3·284	3·115	2·928
9	5·117	4·256	3·863	3·633	3·482	3·374	3·293	3·230	3·137	3·073	2·900	2·707
10	4·965	4·103	3·708	3·478	3·326	3·217	3·135	3·072	2·978	2·913	2·737	2·538
11	4·844	3·982	3·587	3·357	3·204	3·095	3·012	2·948	2·854	2·788	2·609	2·404
12	4·747	3·885	3·490	3·259	3·106	2·996	2·913	2·849	2·753	2·687	2·505	2·296
13	4·667	3·806	3·411	3·179	3·025	2·915	2·832	2·767	2·671	2·604	2·420	2·206
14	4·600	3·739	3·344	3·112	2·958	2·848	2·764	2·699	2·602	2·534	2·349	2·131
15	4·543	3·682	3·287	3·056	2·901	2·790	2·707	2·641	2·544	2·475	2·288	2·066
16	4·494	3·634	3·239	3·007	2·852	2·741	2·657	2·591	2·494	2·425	2·235	2·010
17	4·451	3·592	3·197	2·965	2·810	2·699	2·614	2·548	2·450	2·381	2·190	1·960
18	4·414	3·555	3·160	2·928	2·773	2·661	2·577	2·510	2·412	2·342	2·150	1·917
19	4·381	3·522	3·127	2·895	2·740	2·628	2·544	2·477	2·378	2·308	2·114	1·878
20	4·351	3·493	3·098	2·866	2·711	2·599	2·514	2·447	2·348	2·278	2·082	1·843
21	4·325	3·467	3·072	2·840	2·685	2·573	2·488	2·420	2·321	2·250	2·054	1·812
22	4·301	3·443	3·049	2·817	2·661	2·549	2·464	2·397	2·297	2·226	2·028	1·783
23	4·279	3·422	3·028	2·796	2·640	2·528	2·442	2·375	2·275	2·204	2·005	1·757
24	4·260	3·403	3·009	2·776	2·621	2·508	2·423	2·355	2·255	2·183	1·984	1·733
25	4·242	3·385	2·991	2·759	2·603	2·490	2·405	2·337	2·236	2·165	1·964	1·711
26	4·225	3·369	2·975	2·743	2·587	2·474	2·388	2·321	2·220	2·148	1·946	1·691
27	4·210	3·354	2·960	2·728	2·572	2·459	2·373	2·305	2·204	2·132	1·930	1·672
28	4·196	3·340	2·947	2·714	2·558	2·445	2·359	2·291	2·190	2·118	1·915	1·654
29	4·183	3·328	2·934	2·701	2·545	2·432	2·346	2·278	2·177	2·104	1·901	1·638
30	4·171	3·316	2·922	2·690	2·534	2·421	2·334	2·266	2·165	2·092	1·887	1·622
32	4·149	3·295	2·901	2·668	2·512	2·399	2·313	2·244	2·142	2·070	1·864	1·594
34	4·130	3·276	2·883	2·650	2·494	2·380	2·294	2·225	2·123	2·050	1·843	1·569
36	4·113	3·259	2·866	2·634	2·477	2·364	2·277	2·209	2·106	2·033	1·824	1·547
38	4·098	3·245	2·852	2·619	2·463	2·349	2·262	2·194	2·091	2·017	1·808	1·527
40	4·085	3·232	2·839	2·606	2·449	2·336	2·249	2·180	2·077	2·003	1·793	1·509
60	4·001	3·150	2·758	2·525	2·368	2·254	2·167	2·097	1·993	1·917	1·700	1·389
120	3·920	3·072	2·680	2·447	2·290	2·175	2·087	2·016	1·910	1·834	1·608	1·254
∞	3·841	2·996	2·605	2·372	2·214	2·099	2·010	1·938	1·831	1·752	1·517	1·000

TABLE 12(c). 2·5 PER CENT POINTS OF THE *F*-DISTRIBUTION

The function tabulated is $F(P) = F(P|\nu_1, \nu_2)$ defined by the equation

$$\frac{P}{100} = \frac{\Gamma(\frac{1}{2}\nu_1 + \frac{1}{2}\nu_2)}{\Gamma(\frac{1}{2}\nu_1)\,\Gamma(\frac{1}{2}\nu_2)}\, \nu_1^{\frac{1}{2}\nu_1}\, \nu_2^{\frac{1}{2}\nu_2} \int_{F(P)}^{\infty} \frac{F^{\frac{1}{2}\nu_1 - 1}}{(\nu_2 + \nu_1 F)^{\frac{1}{2}(\nu_1 + \nu_2)}}\, dF,$$

for $P = $ 10, 5, 2·5, 1, 0·5 and 0·1. The lower percentage points, that is the values $F'(P) = F'(P|\nu_1, \nu_2)$ such that the probability that $F \leqslant F'(P)$ is equal to $P/100$, may be found by the formula

$$F'(P|\nu_1, \nu_2) = 1/F(P|\nu_2, \nu_1).$$

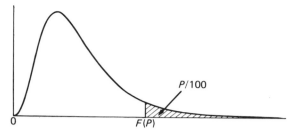

(This shape applies only when $\nu_1 \geqslant 3$. When $\nu_1 < 3$ the mode is at the origin.)

$\nu_1 =$	1	2	3	4	5	6	7	8	10	12	24	∞
$\nu_2 = 1$	647·8	799·5	864·2	899·6	921·8	937·1	948·2	956·7	968·6	976·7	997·2	1018
2	38·51	39·00	39·17	39·25	39·30	39·33	39·36	39·37	39·40	39·41	39·46	39·50
3	17·44	16·04	15·44	15·10	14·88	14·73	14·62	14·54	14·42	14·34	14·12	13·90
4	12·22	10·65	9·979	9·605	9·364	9·197	9·074	8·980	8·844	8·751	8·511	8·257
5	10·01	8·434	7·764	7·388	7·146	6·978	6·853	6·757	6·619	6·525	6·278	6·015
6	8·813	7·260	6·599	6·227	5·988	5·820	5·695	5·600	5·461	5·366	5·117	4·849
7	8·073	6·542	5·890	5·523	5·285	5·119	4·995	4·899	4·761	4·666	4·415	4·142
8	7·571	6·059	5·416	5·053	4·817	4·652	4·529	4·433	4·295	4·200	3·947	3·670
9	7·209	5·715	5·078	4·718	4·484	4·320	4·197	4·102	3·964	3·868	3·614	3·333
10	6·937	5·456	4·826	4·468	4·236	4·072	3·950	3·855	3·717	3·621	3·365	3·080
11	6·724	5·256	4·630	4·275	4·044	3·881	3·759	3·664	3·526	3·430	3·173	2·883
12	6·554	5·096	4·474	4·121	3·891	3·728	3·607	3·512	3·374	3·277	3·019	2·725
13	6·414	4·965	4·347	3·996	3·767	3·604	3·483	3·388	3·250	3·153	2·893	2·595
14	6·298	4·857	4·242	3·892	3·663	3·501	3·380	3·285	3·147	3·050	2·789	2·487
15	6·200	4·765	4·153	3·804	3·576	3·415	3·293	3·199	3·060	2·963	2·701	2·395
16	6·115	4·687	4·077	3·729	3·502	3·341	3·219	3·125	2·986	2·889	2·625	2·316
17	6·042	4·619	4·011	3·665	3·438	3·277	3·156	3·061	2·922	2·825	2·560	2·247
18	5·978	4·560	3·954	3·608	3·382	3·221	3·100	3·005	2·866	2·769	2·503	2·187
19	5·922	4·508	3·903	3·559	3·333	3·172	3·051	2·956	2·817	2·720	2·452	2·133
20	5·871	4·461	3·859	3·515	3·289	3·128	3·007	2·913	2·774	2·676	2·408	2·085
21	5·827	4·420	3·819	3·475	3·250	3·090	2·969	2·874	2·735	2·637	2·368	2·042
22	5·786	4·383	3·783	3·440	3·215	3·055	2·934	2·839	2·700	2·602	2·331	2·003
23	5·750	4·349	3·750	3·408	3·183	3·023	2·902	2·808	2·668	2·570	2·299	1·968
24	5·717	4·319	3·721	3·379	3·155	2·995	2·874	2·779	2·640	2·541	2·269	1·935
25	5·686	4·291	3·694	3·353	3·129	2·969	2·848	2·753	2·613	2·515	2·242	1·906
26	5·659	4·265	3·670	3·329	3·105	2·945	2·824	2·729	2·590	2·491	2·217	1·878
27	5·633	4·242	3·647	3·307	3·083	2·923	2·802	2·707	2·568	2·469	2·195	1·853
28	5·610	4·221	3·626	3·286	3·063	2·903	2·782	2·687	2·547	2·448	2·174	1·829
29	5·588	4·201	3·607	3·267	3·044	2·884	2·763	2·669	2·529	2·430	2·154	1·807
30	5·568	4·182	3·589	3·250	3·026	2·867	2·746	2·651	2·511	2·412	2·136	1·787
32	5·531	4·149	3·557	3·218	2·995	2·836	2·715	2·620	2·480	2·381	2·103	1·750
34	5·499	4·120	3·529	3·191	2·968	2·808	2·688	2·593	2·453	2·353	2·075	1·717
36	5·471	4·094	3·505	3·167	2·944	2·785	2·664	2·569	2·429	2·329	2·049	1·687
38	5·446	4·071	3·483	3·145	2·923	2·763	2·643	2·548	2·407	2·307	2·027	1·661
40	5·424	4·051	3·463	3·126	2·904	2·744	2·624	2·529	2·388	2·288	2·007	1·637
60	5·286	3·925	3·343	3·008	2·786	2·627	2·507	2·412	2·270	2·169	1·882	1·482
120	5·152	3·805	3·227	2·894	2·674	2·515	2·395	2·299	2·157	2·055	1·760	1·310
∞	5·024	3·689	3·116	2·786	2·567	2·408	2·288	2·192	2·048	1·945	1·640	1·000

TABLE 12(d). 1 PER CENT POINTS OF THE F-DISTRIBUTION

If $F = \dfrac{X_1}{\nu_1} \bigg/ \dfrac{X_2}{\nu_2}$, where X_1 and X_2 are independent random variables distributed as χ^2 with ν_1 and ν_2 degrees of freedom respectively, then the probabilities that $F \geqslant F(P)$ and that $F \leqslant F'(P)$ are both equal to $P/100$. Linear interpolation in ν_1 or ν_2 will generally be sufficiently accurate except when either $\nu_1 > 12$ or $\nu_2 > 40$, when harmonic interpolation should be used.

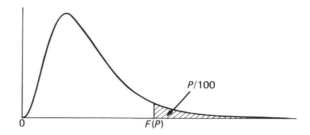

(This shape applies only when $\nu_1 \geqslant 3$. When $\nu_1 < 3$ the mode is at the origin.)

$\nu_1 =$	1	2	3	4	5	6	7	8	10	12	24	∞
$\nu_2 = 1$	4052	4999	5403	5625	5764	5859	5928	5981	6056	6106	6235	6366
2	98·50	99·00	99·17	99·25	99·30	99·33	99·36	99·37	99·40	99·42	99·46	99·50
3	34·12	30·82	29·46	28·71	28·24	27·91	27·67	27·49	27·23	27·05	26·60	26·13
4	21·20	18·00	16·69	15·98	15·52	15·21	14·98	14·80	14·55	14·37	13·93	13·46
5	16·26	13·27	12·06	11·39	10·97	10·67	10·46	10·29	10·05	9·888	9·466	9·020
6	13·75	10·92	9·780	9·148	8·746	8·466	8·260	8·102	7·874	7·718	7·313	6·880
7	12·25	9·547	8·451	7·847	7·460	7·191	6·993	6·840	6·620	6·469	6·074	5·650
8	11·26	8·649	7·591	7·006	6·632	6·371	6·178	6·029	5·814	5·667	5·279	4·859
9	10·56	8·022	6·992	6·422	6·057	5·802	5·613	5·467	5·257	5·111	4·729	4·311
10	10·04	7·559	6·552	5·994	5·636	5·386	5·200	5·057	4·849	4·706	4·327	3·909
11	9·646	7·206	6·217	5·668	5·316	5·069	4·886	4·744	4·539	4·397	4·021	3·602
12	9·330	6·927	5·953	5·412	5·064	4·821	4·640	4·499	4·296	4·155	3·780	3·361
13	9·074	6·701	5·739	5·205	4·862	4·620	4·441	4·302	4·100	3·960	3·587	3·165
14	8·862	6·515	5·564	5·035	4·695	4·456	4·278	4·140	3·939	3·800	3·427	3·004
15	8·683	6·359	5·417	4·893	4·556	4·318	4·142	4·004	3·805	3·666	3·294	2·868
16	8·531	6·226	5·292	4·773	4·437	4·202	4·026	3·890	3·691	3·553	3·181	2·753
17	8·400	6·112	5·185	4·669	4·336	4·102	3·927	3·791	3·593	3·455	3·084	2·653
18	8·285	6·013	5·092	4·579	4·248	4·015	3·841	3·705	3·508	3·371	2·999	2·566
19	8·185	5·926	5·010	4·500	4·171	3·939	3·765	3·631	3·434	3·297	2·925	2·489
20	8·096	5·849	4·938	4·431	4·103	3·871	3·699	3·564	3·368	3·231	2·859	2·421
21	8·017	5·780	4·874	4·369	4·042	3·812	3·640	3·506	3·310	3·173	2·801	2·360
22	7·945	5·719	4·817	4·313	3·988	3·758	3·587	3·453	3·258	3·121	2·749	2·305
23	7·881	5·664	4·765	4·264	3·939	3·710	3·539	3·406	3·211	3·074	2·702	2·256
24	7·823	5·614	4·718	4·218	3·895	3·667	3·496	3·363	3·168	3·032	2·659	2·211
25	7·770	5·568	4·675	4·177	3·855	3·627	3·457	3·324	3·129	2·993	2·620	2·169
26	7·721	5·526	4·637	4·140	3·818	3·591	3·421	3·288	3·094	2·958	2·585	2·131
27	7·677	5·488	4·601	4·106	3·785	3·558	3·388	3·256	3·062	2·926	2·552	2·097
28	7·636	5·453	4·568	4·074	3·754	3·528	3·358	3·226	3·032	2·896	2·522	2·064
29	7·598	5·420	4·538	4·045	3·725	3·499	3·330	3·198	3·005	2·868	2·495	2·034
30	7·562	5·390	4·510	4·018	3·699	3·473	3·304	3·173	2·979	2·843	2·469	2·006
32	7·499	5·336	4·459	3·969	3·652	3·427	3·258	3·127	2·934	2·798	2·423	1·956
34	7·444	5·289	4·416	3·927	3·611	3·386	3·218	3·087	2·894	2·758	2·383	1·911
36	7·396	5·248	4·377	3·890	3·574	3·351	3·183	3·052	2·859	2·723	2·347	1·872
38	7·353	5·211	4·343	3·858	3·542	3·319	3·152	3·021	2·828	2·692	2·316	1·837
40	7·314	5·179	4·313	3·828	3·514	3·291	3·124	2·993	2·801	2·665	2·288	1·805
60	7·077	4·977	4·126	3·649	3·339	3·119	2·953	2·823	2·632	2·496	2·115	1·601
120	6·851	4·787	3·949	3·480	3·174	2·956	2·792	2·663	2·472	2·336	1·950	1·381
∞	6·635	4·605	3·782	3·319	3·017	2·802	2·639	2·511	2·321	2·185	1·791	1·000

TABLE 12(e). 0·5 PER CENT POINTS OF THE *F*-DISTRIBUTION

The function tabulated is $F(P) = F(P|\nu_1, \nu_2)$ defined by the equation

$$\frac{P}{100} = \frac{\Gamma(\tfrac{1}{2}\nu_1 + \tfrac{1}{2}\nu_2)}{\Gamma(\tfrac{1}{2}\nu_1)\,\Gamma(\tfrac{1}{2}\nu_2)}\,\nu_1^{\frac{1}{2}\nu_1}\,\nu_2^{\frac{1}{2}\nu_2}\int_{F(P)}^{\infty}\frac{F^{\frac{1}{2}\nu_1-1}}{(\nu_2+\nu_1 F)^{\frac{1}{2}(\nu_1+\nu_2)}}\,dF,$$

for $P = 10, 5, 2\cdot5, 1, 0\cdot5$ and $0\cdot1$. The lower percentage points, that is the values $F'(P) = F'(P|\nu_1, \nu_2)$ such that the probability that $F \leqslant F'(P)$ is equal to $P/100$, may be found by the formula

$$F'(P|\nu_1, \nu_2) = 1/F(P|\nu_2, \nu_1).$$

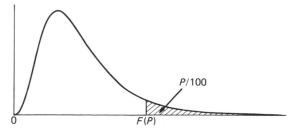

(This shape applies only when $\nu_1 \geqslant 3$. When $\nu_1 < 3$ the mode is at the origin.)

$\nu_1 =$	1	2	3	4	5	6	7	8	10	12	24	∞
$\nu_2 = 1$	16211	20000	21615	22500	23056	23437	23715	23925	24224	24426	24940	25464
2	198·5	199·0	199·2	199·2	199·3	199·3	199·4	199·4	199·4	199·4	199·5	199·5
3	55·55	49·80	47·47	46·19	45·39	44·84	44·43	44·13	43·69	43·39	42·62	41·83
4	31·33	26·28	24·26	23·15	22·46	21·97	21·62	21·35	20·97	20·70	20·03	19·32
5	22·78	18·31	16·53	15·56	14·94	14·51	14·20	13·96	13·62	13·38	12·78	12·14
6	18·63	14·54	12·92	12·03	11·46	11·07	10·79	10·57	10·25	10·03	9·474	8·879
7	16·24	12·40	10·88	10·05	9·522	9·155	8·885	8·678	8·380	8·176	7·645	7·076
8	14·69	11·04	9·596	8·805	8·302	7·952	7·694	7·496	7·211	7·015	6·503	5·951
9	13·61	10·11	8·717	7·956	7·471	7·134	6·885	6·693	6·417	6·227	5·729	5·188
10	12·83	9·427	8·081	7·343	6·872	6·545	6·302	6·116	5·847	5·661	5·173	4·639
11	12·23	8·912	7·600	6·881	6·422	6·102	5·865	5·682	5·418	5·236	4·756	4·226
12	11·75	8·510	7·226	6·521	6·071	5·757	5·525	5·345	5·085	4·906	4·431	3·904
13	11·37	8·186	6·926	6·233	5·791	5·482	5·253	5·076	4·820	4·643	4·173	3·647
14	11·06	7·922	6·680	5·998	5·562	5·257	5·031	4·857	4·603	4·428	3·961	3·436
15	10·80	7·701	6·476	5·803	5·372	5·071	4·847	4·674	4·424	4·250	3·786	3·260
16	10·58	7·514	6·303	5·638	5·212	4·913	4·692	4·521	4·272	4·099	3·638	3·112
17	10·38	7·354	6·156	5·497	5·075	4·779	4·559	4·389	4·142	3·971	3·511	2·984
18	10·22	7·215	6·028	5·375	4·956	4·663	4·445	4·276	4·030	3·860	3·402	2·873
19	10·07	7·093	5·916	5·268	4·853	4·561	4·345	4·177	3·933	3·763	3·306	2·776
20	9·944	6·986	5·818	5·174	4·762	4·472	4·257	4·090	3·847	3·678	3·222	2·690
21	9·830	6·891	5·730	5·091	4·681	4·393	4·179	4·013	3·771	3·602	3·147	2·614
22	9·727	6·806	5·652	5·017	4·609	4·322	4·109	3·944	3·703	3·535	3·081	2·545
23	9·635	6·730	5·582	4·950	4·544	4·259	4·047	3·882	3·642	3·475	3·021	2·484
24	9·551	6·661	5·519	4·890	4·486	4·202	3·991	3·826	3·587	3·420	2·967	2·428
25	9·475	6·598	5·462	4·835	4·433	4·150	3·939	3·776	3·537	3·370	2·918	2·377
26	9·406	6·541	5·409	4·785	4·384	4·103	3·893	3·730	3·492	3·325	2·873	2·330
27	9·342	6·489	5·361	4·740	4·340	4·059	3·850	3·687	3·450	3·284	2·832	2·287
28	9·284	6·440	5·317	4·698	4·300	4·020	3·811	3·649	3·412	3·246	2·794	2·247
29	9·230	6·396	5·276	4·659	4·262	3·983	3·775	3·613	3·377	3·211	2·759	2·210
30	9·180	6·355	5·239	4·623	4·228	3·949	3·742	3·580	3·344	3·179	2·727	2·176
32	9·090	6·281	5·171	4·559	4·166	3·889	3·682	3·521	3·286	3·121	2·670	2·114
34	9·012	6·217	5·113	4·504	4·112	3·836	3·630	3·470	3·235	3·071	2·620	2·060
36	8·943	6·161	5·062	4·455	4·065	3·790	3·585	3·425	3·191	3·027	2·576	2·013
38	8·882	6·111	5·016	4·412	4·023	3·749	3·545	3·385	3·152	2·988	2·537	1·970
40	8·828	6·066	4·976	4·374	3·986	3·713	3·509	3·350	3·117	2·953	2·502	1·932
60	8·495	5·795	4·729	4·140	3·760	3·492	3·291	3·134	2·904	2·742	2·290	1·689
120	8·179	5·539	4·497	3·921	3·548	3·285	3·087	2·933	2·705	2·544	2·089	1·431
∞	7·879	5·298	4·279	3·715	3·350	3·091	2·897	2·744	2·519	2·358	1·898	1·000

TABLE 12(f). 0·1 PER CENT POINTS OF THE F-DISTRIBUTION

If $F = \dfrac{X_1}{\nu_1} \Big/ \dfrac{X_2}{\nu_2}$, where X_1 and X_2 are independent random variables distributed as χ^2 with ν_1 and ν_2 degrees of freedom respectively, then the probabilities that $F \geqslant F(P)$ and that $F \leqslant F'(P)$ are both equal to $P/100$. Linear interpolation in ν_1 or ν_2 will generally be sufficiently accurate except when either $\nu_1 > 12$ or $\nu_2 > 40$, when harmonic interpolation should be used.

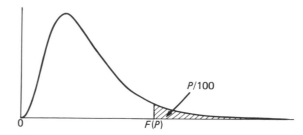

(This shape applies only when $\nu_1 \geqslant 3$. When $\nu_1 < 3$ the mode is at the origin.)

$\nu_1 =$	1	2	3	4	5	6	7	8	10	12	24	∞
$\nu_2 = 1^*$	4053	5000	5404	5625	5764	5859	5929	5981	6056	6107	6235	6366
2	998·5	999·0	999·2	999·2	999·3	999·3	999·4	999·4	999·4	999·4	999·5	999·5
3	167·0	148·5	141·1	137·1	134·6	132·8	131·6	130·6	129·2	128·3	125·9	123·5
4	74·14	61·25	56·18	53·44	51·71	50·53	49·66	49·00	48·05	47·41	45·77	44·05
5	47·18	37·12	33·20	31·09	29·75	28·83	28·16	27·65	26·92	26·42	25·13	23·79
6	35·51	27·00	23·70	21·92	20·80	20·03	19·46	19·03	18·41	17·99	16·90	15·75
7	29·25	21·69	18·77	17·20	16·21	15·52	15·02	14·63	14·08	13·71	12·73	11·70
8	25·41	18·49	15·83	14·39	13·48	12·86	12·40	12·05	11·54	11·19	10·30	9·334
9	22·86	16·39	13·90	12·56	11·71	11·13	10·70	10·37	9·894	9·570	8·724	7·813
10	21·04	14·91	12·55	11·28	10·48	9·926	9·517	9·204	8·754	8·445	7·638	6·762
11	19·69	13·81	11·56	10·35	9·578	9·047	8·655	8·355	7·922	7·626	6·847	5·998
12	18·64	12·97	10·80	9·633	8·892	8·379	8·001	7·710	7·292	7·005	6·249	5·420
13	17·82	12·31	10·21	9·073	8·354	7·856	7·489	7·206	6·799	6·519	5·781	4·967
14	17·14	11·78	9·729	8·622	7·922	7·436	7·077	6·802	6·404	6·130	5·407	4·604
15	16·59	11·34	9·335	8·253	7·567	7·092	6·741	6·471	6·081	5·812	5·101	4·307
16	16·12	10·97	9·006	7·944	7·272	6·805	6·460	6·195	5·812	5·547	4·846	4·059
17	15·72	10·66	8·727	7·683	7·022	6·562	6·223	5·962	5·584	5·324	4·631	3·850
18	15·38	10·39	8·487	7·459	6·808	6·355	6·021	5·763	5·390	5·132	4·447	3·670
19	15·08	10·16	8·280	7·265	6·622	6·175	5·845	5·590	5·222	4·967	4·288	3·514
20	14·82	9·953	8·098	7·096	6·461	6·019	5·692	5·440	5·075	4·823	4·149	3·378
21	14·59	9·772	7·938	6·947	6·318	5·881	5·557	5·308	4·946	4·696	4·027	3·257
22	14·38	9·612	7·796	6·814	6·191	5·758	5·438	5·190	4·832	4·583	3·919	3·151
23	14·20	9·469	7·669	6·696	6·078	5·649	5·331	5·085	4·730	4·483	3·822	3·055
24	14·03	9·339	7·554	6·589	5·977	5·550	5·235	4·991	4·638	4·393	3·735	2·969
25	13·88	9·223	7·451	6·493	5·885	5·462	5·148	4·906	4·555	4·312	3·657	2·890
26	13·74	9·116	7·357	6·406	5·802	5·381	5·070	4·829	4·480	4·238	3·586	2·819
27	13·61	9·019	7·272	6·326	5·726	5·308	4·998	4·759	4·412	4·171	3·521	2·754
28	13·50	8·931	7·193	6·253	5·656	5·241	4·933	4·695	4·349	4·109	3·462	2·695
29	13·39	8·849	7·121	6·186	5·593	5·179	4·873	4·636	4·292	4·053	3·407	2·640
30	13·29	8·773	7·054	6·125	5·534	5·122	4·817	4·581	4·239	4·001	3·357	2·589
32	13·12	8·639	6·936	6·014	5·429	5·021	4·719	4·485	4·145	3·908	3·268	2·498
34	12·97	8·522	6·833	5·919	5·339	4·934	4·633	4·401	4·063	3·828	3·191	2·419
36	12·83	8·420	6·744	5·836	5·260	4·857	4·559	4·328	3·992	3·758	3·123	2·349
38	12·71	8·331	6·665	5·763	5·190	4·790	4·494	4·264	3·930	3·697	3·064	2·288
40	12·61	8·251	6·595	5·698	5·128	4·731	4·436	4·207	3·874	3·642	3·011	2·233
60	11·97	7·768	6·171	5·307	4·757	4·372	4·086	3·865	3·541	3·315	2·694	1·890
120	11·38	7·321	5·781	4·947	4·416	4·044	3·767	3·552	3·237	3·016	2·402	1·543
∞	10·83	6·908	5·422	4·617	4·103	3·743	3·475	3·266	2·959	2·742	2·132	1·000

* Entries in the row $\nu_2 = 1$ must be multiplied by 100.

TABLE 13. PERCENTAGE POINTS OF THE CORRELATION COEFFICIENT r WHEN $\rho = 0$

The function tabulated is $r(P) = r(P|\nu)$ defined by the equation

$$\frac{\Gamma\left(\frac{\nu-1}{2}\right)}{\sqrt{\pi}\,\Gamma\left(\frac{\nu-2}{2}\right)} \int_{r(P)}^{1} (1-r^2)^{\frac{\nu-4}{2}}\,dr = P/100.$$

Let r be a partial correlation coefficient, after s variables have been eliminated, in a sample of size n from a multivariate normal population with corresponding true partial correlation coefficient $\rho = 0$, and let $\nu = n-s$. This table gives upper P per cent points of r; the corresponding lower P per cent points are given by $-r(P)$, and the tabulated values are also upper $2P$ per cent points of $|r|$. For $s = 0$ we have $\nu = n$ and r is the ordinary correlation coefficient. When $\nu > 130$ use the results that r is approximately normally distributed with zero mean and variance $\dfrac{1}{\nu-1}$, or (more accurately) that $z = \tanh^{-1} r$ is approximately normally distributed with zero mean and variance $\dfrac{1}{\nu-3}$ (cf. Tables 16 and 17).

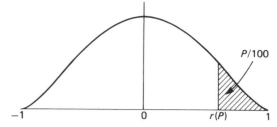

(This shape applies for $\nu \geqslant 5$ only. When $\nu = 4$ the distribution is uniform and when $\nu = 3$ the probability density function is U-shaped.)

Tables of the distribution of r for various values of ρ are given by, for example, F. N. David, *Tables of the Ordinates and Probability Integral of the Distribution of the Correlation Coefficient in Small Samples*, Cambridge University Press (1954), and R. E. Odeh, 'Critical values of the sample product–moment correlation coefficient in the bivariate normal distribution', *Commun. Statist. – Simula Computa.* **11** (1) (1982), pp. 1–26. The z-transformation may also be used (cf. Tables 16 and 17).

P	5	2·5	1	0·5	0·1
$\nu = 3$	0·9877	0·9969	0·9995	0·9999	0·999995
4	·9000	·9500	·9800	·9900	·9980
5	0·8054	0·8783	0·9343	0·9587	0·9859
6	·7293	·8114	·8822	·9172	·9633
7	·6694	·7545	·8329	·8745	·9350
8	·6215	·7067	·7887	·8343	·9049
9	·5822	·6664	·7498	·7977	·8751
10	0·5494	0·6319	0·7155	0·7646	0·8467
11	·5214	·6021	·6851	·7348	·8199
12	·4973	·5760	·6581	·7079	·7950
13	·4762	·5529	·6339	·6835	·7717
14	·4575	·5324	·6120	·6614	·7501
15	0·4409	0·5140	0·5923	0·6411	0·7301
16	·4259	·4973	·5742	·6226	·7114
17	·4124	·4821	·5577	·6055	·6940
18	·4000	·4683	·5425	·5897	·6777
19	·3887	·4555	·5285	·5751	·6624
20	0·3783	0·4438	0·5155	0·5614	0·6481
21	·3687	·4329	·5034	·5487	·6346
22	·3598	·4227	·4921	·5368	·6219
23	·3515	·4132	·4815	·5256	·6099
24	·3438	·4044	·4716	·5151	·5986
25	0·3365	0·3961	0·4622	0·5052	0·5879
26	·3297	·3882	·4534	·4958	·5776
27	·3233	·3809	·4451	·4869	·5679
28	·3172	·3739	·4372	·4785	·5587
29	·3115	·3673	·4297	·4705	·5499
30	0·3061	0·3610	0·4226	0·4629	0·5415
31	·3009	·3550	·4158	·4556	·5334
32	·2960	·3494	·4093	·4487	·5257
33	·2913	·3440	·4032	·4421	·5184
34	·2869	·3388	·3972	·4357	·5113
35	0·2826	0·3338	0·3916	0·4296	0·5045
36	·2785	·3291	·3862	·4238	·4979
37	·2746	·3246	·3810	·4182	·4916
38	·2709	·3202	·3760	·4128	·4856
39	·2673	·3160	·3712	·4076	·4797
P	5	2·5	1	0·5	0·1
$\nu = 40$	0·2638	0·3120	0·3665	0·4026	0·4741
42	·2573	·3044	·3578	·3932	·4633
44	·2512	·2973	·3496	·3843	·4533
46	·2455	·2907	·3420	·3761	·4439
48	·2403	·2845	·3348	·3683	·4351
50	0·2353	0·2787	0·3281	0·3610	0·4267
52	·2306	·2732	·3218	·3542	·4188
54	·2262	·2681	·3158	·3477	·4114
56	·2221	·2632	·3102	·3415	·4043
58	·2181	·2586	·3048	·3357	·3976
60	0·2144	0·2542	0·2997	0·3301	0·3912
62	·2108	·2500	·2948	·3248	·3850
64	·2075	·2461	·2902	·3198	·3792
66	·2042	·2423	·2858	·3150	·3736
68	·2012	·2387	·2816	·3104	·3683
70	0·1982	0·2352	0·2776	0·3060	0·3632
72	·1954	·2319	·2737	·3017	·3583
74	·1927	·2287	·2700	·2977	·3536
76	·1901	·2257	·2664	·2938	·3490
78	·1876	·2227	·2630	·2900	·3447
80	0·1852	0·2199	0·2597	0·2864	0·3405
82	·1829	·2172	·2565	·2830	·3364
84	·1807	·2146	·2535	·2796	·3325
86	·1786	·2120	·2505	·2764	·3287
88	·1765	·2096	·2477	·2732	·3251
90	0·1745	0·2072	0·2449	0·2702	0·3215
92	·1726	·2050	·2422	·2673	·3181
94	·1707	·2028	·2396	·2645	·3148
96	·1689	·2006	·2371	·2617	·3116
98	·1671	·1986	·2347	·2591	·3085
100	0·1654	0·1966	0·2324	0·2565	0·3054
105	·1614	·1918	·2268	·2504	·2983
110	·1576	·1874	·2216	·2446	·2915
115	·1541	·1832	·2167	·2393	·2853
120	·1509	·1793	·2122	·2343	·2794
125	0·1478	0·1757	0·2079	0·2296	0·2738
130	·1449	·1723	·2039	·2252	·2686

TABLE 14. PERCENTAGE POINTS OF SPEARMAN'S S
TABLE 15. PERCENTAGE POINTS OF KENDALL'S K

Spearman's S and Kendall's K are both used to measure the degree of association between two rankings of n objects. Let d_i ($1 \leqslant i \leqslant n$) be the difference in the ranks of the ith object; Spearman's S is defined as $\sum_{i=1}^{n} d_i^2$. To define Kendall's K, re-order the pairs of ranks so that the first set is in natural order from left to right, and let m_i ($1 \leqslant i \leqslant n$) be the number of ranks greater than i in the second ranking which are to the right of rank i. Kendall's K is defined as $\sum_{i=1}^{n} m_i$.

For Table 14 the tabulated value $x(P)$ is the lower percentage point, i.e. the largest value x such that, in independent rankings, $\Pr(S \leqslant x) \leqslant P/100$; in Table 15, K replaces S and the upper percentage point is given. A dash indicates that there is no value with the required property. The distributions are symmetric about means $\frac{1}{6}(n^3-n)$ for S and $\frac{1}{4}n(n-1)$ for K, with maxima equal to twice the means; hence the upper percentage points of S are $\frac{1}{6}(n^3-n) - x(P)$ and the lower percentage points of K are $\frac{1}{2}n(n-1) - x(P)$. The variances are $\frac{1}{36}n^2(n+1)^2(n-1)$ for S and $\frac{1}{72}n(n-1)(2n+5)$ for K, and when $n > 40$ both statistics are approximately normally distributed; more accurately, the distribution function of $X = [S - \frac{1}{6}(n^3-n)]/[\frac{1}{6}n(n+1)\sqrt{n-1}]$ is approximately equal to $\Phi(x) - \dfrac{\gamma}{24\sqrt{2\pi}} e^{-\frac{1}{2}x^2}(x^3 - 3x)$, where $\gamma = \dfrac{-0.04(19n^2 + 5n - 36)}{\frac{1}{6}(n^3-n)}$ and $\Phi(x)$ is the normal distribution function (see Table 4).

A test of the null hypothesis of independent rankings is provided by rejecting at the P per cent level if $S \leqslant x(P)$, or $K \geqslant x(P)$, when the alternative is contrary rankings. The other points are similarly used when the alternative is similar rankings. To cover both alternatives reject at the $2P$ per cent level if S, or K, lies in either tail. Spearman's rank correlation coefficient r_S is defined as $1 - 6S/(n^3-n)$, and has upper and lower P per cent points $1 - 6x(P)/(n^3-n)$ and $-[1 - 6x(P)/(n^3-n)]$ respectively. Kendall's rank correlation coefficient r_K is defined as $4K/[n(n-1)] - 1$, and has upper and lower P per cent points $4x(P)/[n(n-1)] - 1$ and $-\{4x(P)/[n(n-1)] - 1\}$ respectively.

SPEARMAN'S S

P	5	2·5	1	0·5	0·1	$\frac{1}{6}(n^3-n)$
$n=4$	0	—	—	—	—	10
5	2	0	0	—	—	20
6	6	4	2	0	—	35
7	16	12	6	4	0	56
8	30	22	14	10	4	84
9	48	36	26	20	10	120
10	72	58	42	34	20	165
11	102	84	64	54	34	220
12	142	118	92	78	52	286
13	188	160	128	108	76	364
14	244	210	170	146	104	455
15	310	268	222	194	140	560
16	388	338	284	248	184	680
17	478	418	354	312	236	816
18	580	512	436	388	298	969
19	694	616	530	474	370	1140
20	824	736	636	572	452	1330
21	970	868	756	684	544	1540
22	1132	1018	890	808	650	1771
23	1310	1182	1040	948	768	2024
24	1508	1364	1206	1102	900	2300
25	1724	1566	1388	1272	1048	2600
26	1958	1784	1588	1460	1210	2925
27	2214	2022	1806	1664	1388	3276
28	2492	2282	2044	1888	1584	3654
29	2794	2564	2304	2132	1796	4060
30	3118	2866	2584	2396	2028	4495
31	3466	3194	2884	2682	2280	4960
32	3840	3544	3210	2988	2552	5456
33	4240	3920	3558	3318	2844	5984
34	4666	4322	3930	3672	3160	6545
35	5120	4750	4330	4050	3498	7140
36	5604	5206	4754	4454	3858	7770
37	6118	5692	5206	4884	4244	8436
38	6662	6206	5686	5342	4656	9139
39	7238	6750	6196	5826	5092	9880
40	7846	7326	6736	6342	5556	10660

KENDALL'S K

P	5	2·5	1	0·5	0·1	$\frac{1}{4}n(n-1)$
$n=4$	6	—	—	—	—	3
5	9	10	10	—	—	5
6	13	14	14	15	—	7·5
7	17	18	19	20	21	10·5
8	22	23	24	25	26	14
9	27	28	30	31	33	18
10	33	34	36	37	40	22·5
11	39	41	43	44	47	27·5
12	46	48	51	52	55	33
13	53	56	59	61	64	39
14	62	64	67	69	73	45·5
15	70	73	77	79	83	52·5
16	79	83	86	89	94	60
17	89	93	97	100	105	68
18	99	103	108	111	117	76·5
19	110	114	119	123	129	85·5
20	121	126	131	135	142	95
21	133	138	144	148	156	105
22	146	151	157	161	170	115·5
23	159	164	171	176	184	126·5
24	172	178	185	190	200	138
25	186	193	200	205	216	150
26	201	208	216	221	232	162·5
27	216	223	232	238	249	175·5
28	232	239	248	254	267	189
29	248	256	266	272	285	203
30	265	273	283	290	303	217·5
31	282	291	301	308	323	232·5
32	300	309	320	328	342	248
33	318	328	340	347	363	264
34	337	347	359	368	384	280·5
35	356	367	380	388	405	297·5
36	376	388	401	410	428	315
37	397	409	422	432	450	333
38	418	430	444	454	473	351·5
39	440	452	467	477	497	370·5
40	462	475	490	501	522	390

TABLE 16. THE z-TRANSFORMATION OF THE CORRELATION COEFFICIENT

The function tabulated is $z = \tanh^{-1} r = \frac{1}{2}\log_e\left(\frac{1+r}{1-r}\right)$. If $r < 0$ use the negative of the value of z for $-r$. Let r be a partial correlation coefficient, after s variables have been eliminated, in a sample of size n from a multivariate normal population with the corresponding true partial correlation coefficient ρ, and let $\nu = n-s$. Then z is approximately normally distributed with mean $\tanh^{-1}\rho + \rho/2(\nu-1)$ (or, less accurately, $\tanh^{-1}\rho$) and variance $1/(\nu-3)$. If $s = 0$ we have $\nu = n$ and r is the ordinary correlation coefficient. For $\rho = 0$ the exact percentage points are given in Table 13.

r	z	r	z	r	z	r	z	r	z	r	z
0·00	0·0000	0·500	0·5493	0·750	0·9730	0·910	1·5275	0·9700	2·0923	0·9950	2·9945
·01	·0100	·505	·5560	·755	0·9845	·912	·5393	·9705	·1008	·9951	3·0046
·02	·0200	·510	·5627	·760	0·9962	·914	·5513	·9710	·1095	·9952	3·0149
·03	·0300	·515	·5695	·765	1·0082	·916	·5636	·9715	·1183	·9953	3·0255
·04	·0400	·520	·5763	·770	1·0203	·918	·5762	·9720	·1273	·9954	3·0363
0·05	0·0500	0·525	0·5832	0·775	1·0327	0·920	1·5890	0·9725	2·1364	0·9955	3·0473
·06	·0601	·530	·5901	·780	·0454	·922	·6022	·9730	·1457	·9956	·0585
·07	·0701	·535	·5971	·785	·0583	·924	·6157	·9735	·1552	·9957	·0701
·08	·0802	·540	·6042	·790	·0714	·926	·6296	·9740	·1649	·9958	·0819
·09	·0902	·545	·6112	·795	·0849	·928	·6438	·9745	·1747	·9959	·0939
0·10	0·1003	0·550	0·6184	0·800	1·0986	0·930	1·6584	0·9750	2·1847	0·9960	3·1063
·11	·1104	·555	·6256	·805	·1127	·931	·6658	·9755	·1950	·9961	·1190
·12	·1206	·560	·6328	·810	·1270	·932	·6734	·9760	·2054	·9962	·1320
·13	·1307	·565	·6401	·815	·1417	·933	·6811	·9765	·2160	·9963	·1454
·14	·1409	·570	·6475	·820	·1568	·934	·6888	·9770	·2269	·9964	·1591
0·15	0·1511	0·575	0·6550	0·825	1·1723	0·935	1·6967	0·9775	2·2380	0·9965	3·1732
·16	·1614	·580	·6625	·830	·1881	·936	·7047	·9780	·2494	·9966	·1877
·17	·1717	·585	·6700	·835	·2044	·937	·7129	·9785	·2610	·9967	·2027
·18	·1820	·590	·6777	·840	·2212	·938	·7211	·9790	·2729	·9968	·2181
·19	·1923	·595	·6854	·845	·2384	·939	·7295	·9795	·2851	·9969	·2340
0·20	0·2027	0·600	0·6931	0·850	1·2562	0·940	1·7380	0·9800	2·2976	0·9970	3·2504
·21	·2132	·605	·7010	·852	·2634	·941	·7467	·9805	·3103	·9971	·2674
·22	·2237	·610	·7089	·854	·2707	·942	·7555	·9810	·3235	·9972	·2849
·23	·2342	·615	·7169	·856	·2782	·943	·7645	·9815	·3369	·9973	·3031
·24	·2448	·620	·7250	·858	·2857	·944	·7736	·9820	·3507	·9974	·3220
0·25	0·2554	0·625	0·7332	0·860	1·2933	0·945	1·7828	0·9825	2·3650	0·9975	3·3417
·26	·2661	·630	·7414	·862	·3011	·946	·7923	·9830	·3796	·9976	·3621
·27	·2769	·635	·7498	·864	·3089	·947	·8019	·9835	·3946	·9977	·3834
·28	·2877	·640	·7582	·866	·3169	·948	·8117	·9840	·4101	·9978	·4057
·29	·2986	·645	·7667	·868	·3249	·949	·8216	·9845	·4261	·9979	·4290
0·30	0·3095	0·650	0·7753	0·870	1·3331	0·950	1·8318	0·9850	2·4427	0·9980	3·4534
·31	·3205	·655	·7840	·872	·3414	·951	·8421	·9855	·4597	·9981	·4790
·32	·3316	·660	·7928	·874	·3498	·952	·8527	·9860	·4774	·9982	·5061
·33	·3428	·665	·8017	·876	·3583	·953	·8635	·9865	·4957	·9983	·5347
·34	·3541	·670	·8107	·878	·3670	·954	·8745	·9870	·5147	·9984	·5650
0·35	0·3654	0·675	0·8199	0·880	1·3758	0·955	1·8857	0·9875	2·5345	0·9985	3·5973
·36	·3769	·680	·8291	·882	·3847	·956	·8972	·9880	·5550	·9986	·6319
·37	·3884	·685	·8385	·884	·3938	·957	·9090	·9885	·5764	·9987	·6689
·38	·4001	·690	·8480	·886	·4030	·958	·9210	·9890	·5987	·9988	·7090
·39	·4118	·695	·8576	·888	·4124	·959	·9333	·9895	·6221	·9989	·7525
0·40	0·4236	0·700	0·8673	0·890	1·4219	0·960	1·9459	0·9900	2·6467	0·9990	3·8002
·41	·4356	·705	·8772	·892	·4316	·961	·9588	·9905	·6724	·9991	3·8529
·42	·4477	·710	·8872	·894	·4415	·962	·9721	·9910	·6996	·9992	3·9118
·43	·4599	·715	·8973	·896	·4516	·963	·9857	·9915	·7283	·9993	3·9786
·44	·4722	·720	·9076	·898	·4618	·964	·9996	·9920	·7587	·9994	4·0557
0·45	0·4847	0·725	0·9181	0·900	1·4722	0·965	2·0139	0·9925	2·7911	0·9995	4·1469
·46	·4973	·730	·9287	·902	·4828	·966	·0287	·9930	·8257	·9996	·2585
·47	·5101	·735	·9395	·904	·4937	·967	·0439	·9935	·8629	·9997	·4024
·48	·5230	·740	·9505	·906	·5047	·968	·0595	·9940	·9031	·9998	·6051
·49	·5361	·745	·9616	·908	·5160	·969	·0756	·9945	·9467	·9999	·9517
0·50	0·5493	0·750	0·9730	0·910	1·5275	0·970	2·0923	0·9950	2·9945	1·0000	∞

TABLE 17. THE INVERSE OF THE z-TRANSFORMATION

The function tabulated is $r = \tanh z = \dfrac{e^{2z}-1}{e^{2z}+1}$. If $z < 0$, use the negative of the value of r for $-z$.

z	r	z	r	z	r	z	r	z	r	z	r
0·00	0·0000	0·50	0·4621	1·00	0·7616	1·50	0·9051	2·00	0·9640	3·00	0·9951
·01	·0100	·51	·4699	·01	·7658	·51	·9069	·02	·9654	·02	·9952
·02	·0200	·52	·4777	·02	·7699	·52	·9087	·04	·9667	·04	·9954
·03	·0300	·53	·4854	·03	·7739	·53	·9104	·06	·9680	·06	·9956
·04	·0400	·54	·4930	·04	·7779	·54	·9121	·08	·9693	·08	·9958
0·05	0·0500	0·55	0·5005	1·05	0·7818	1·55	0·9138	2·10	0·9705	3·10	0·9959
·06	·0599	·56	·5080	·06	·7857	·56	·9154	·12	·9716	·12	·9961
·07	·0699	·57	·5154	·07	·7895	·57	·9170	·14	·9727	·14	·9963
·08	·0798	·58	·5227	·08	·7932	·58	·9186	·16	·9737	·16	·9964
·09	·0898	·59	·5299	·09	·7969	·59	·9201	·18	·9748	·18	·9965
0·10	0·0997	0·60	0·5370	1·10	0·8005	1·60	0·9217	2·20	0·9757	3·20	0·9967
·11	·1096	·61	·5441	·11	·8041	·61	·9232	·22	·9767	·22	·9968
·12	·1194	·62	·5511	·12	·8076	·62	·9246	·24	·9776	·24	·9969
·13	·1293	·63	·5581	·13	·8110	·63	·9261	·26	·9785	·26	·9971
·14	·1391	·64	·5649	·14	·8144	·64	·9275	·28	·9793	·28	·9972
0·15	0·1489	0·65	0·5717	1·15	0·8178	1·65	0·9289	2·30	0·9801	3·30	0·9973
·16	·1586	·66	·5784	·16	·8210	·66	·9302	·32	·9809	·32	·9974
·17	·1684	·67	·5850	·17	·8243	·67	·9316	·34	·9816	·34	·9975
·18	·1781	·68	·5915	·18	·8275	·68	·9329	·36	·9823	·36	·9976
·19	·1877	·69	·5980	·19	·8306	·69	·9341	·38	·9830	·38	·9977
0·20	0·1974	0·70	0·6044	1·20	0·8337	1·70	0·9354	2·40	0·9837	3·40	0·9978
·21	·2070	·71	·6107	·21	·8367	·71	·9366	·42	·9843	·42	·9979
·22	·2165	·72	·6169	·22	·8397	·72	·9379	·44	·9849	·44	·9979
·23	·2260	·73	·6231	·23	·8426	·73	·9391	·46	·9855	·46	·9980
·24	·2355	·74	·6291	·24	·8455	·74	·9402	·48	·9861	·48	·9981
0·25	0·2449	0·75	0·6351	1·25	0·8483	1·75	0·9414	2·50	0·9866	3·50	0·9982
·26	·2543	·76	·6411	·26	·8511	·76	·9425	·52	·9871	·55	·9984
·27	·2636	·77	·6469	·27	·8538	·77	·9436	·54	·9876	·60	·9985
·28	·2729	·78	·6527	·28	·8565	·78	·9447	·56	·9881	·65	·9986
·29	·2821	·79	·6584	·29	·8591	·79	·9458	·58	·9886	·70	·9988
0·30	0·2913	0·80	0·6640	1·30	0·8617	1·80	0·9468	2·60	0·9890	3·75	0·9989
·31	·3004	·81	·6696	·31	·8643	·81	·9478	·62	·9895	·80	·9990
·32	·3095	·82	·6751	·32	·8668	·82	·9488	·64	·9899	·85	·9991
·33	·3185	·83	·6805	·33	·8692	·83	·9498	·66	·9903	·90	·9992
·34	·3275	·84	·6858	·34	·8717	·84	·9508	·68	·9906	·95	·9993
0·35	0·3364	0·85	0·6911	1·35	0·8741	1·85	0·9517	2·70	0·9910	4·00	0·9993
·36	·3452	·86	·6963	·36	·8764	·86	·9527	·72	·9914	·05	·9994
·37	·3540	·87	·7014	·37	·8787	·87	·9536	·74	·9917	·10	·9995
·38	·3627	·88	·7064	·38	·8810	·88	·9545	·76	·9920	·15	·9995
·39	·3714	·89	·7114	·39	·8832	·89	·9554	·78	·9923	·20	·9996
0·40	0·3799	0·90	0·7163	1·40	0·8854	1·90	0·9562	2·80	0·9926	4·25	0·9996
·41	·3885	·91	·7211	·41	·8875	·91	·9571	·82	·9929	·30	·9996
·42	·3969	·92	·7259	·42	·8896	·92	·9579	·84	·9932	·35	·9997
·43	·4053	·93	·7306	·43	·8917	·93	·9587	·86	·9935	·40	·9997
·44	·4136	·94	·7352	·44	·8937	·94	·9595	·88	·9937	·45	·9997
0·45	0·4219	0·95	0·7398	1·45	0·8957	1·95	0·9603	2·90	0·9940	4·50	0·9998
·46	·4301	·96	·7443	·46	·8977	·96	·9611	·92	·9942	·55	·9998
·47	·4382	·97	·7487	·47	·8996	·97	·9618	·94	·9944	·60	·9998
·48	·4462	·98	·7531	·48	·9015	·98	·9626	·96	·9946	·65	·9998
·49	·4542	·99	·7574	·49	·9033	·99	·9633	·98	·9949	·70	·9998
0·50	0·4621	1·00	0·7616	1·50	0·9051	2·00	0·9640	3·00	0·9951	4·75	0·9999

TABLE 18. PERCENTAGE POINTS OF THE DISTRIBUTION OF THE NUMBER OF RUNS

Suppose that n_1 A's and n_2 B's ($n_1 \leq n_2$) are arranged at random in a row, and let R be the number of runs (that is, sets of one or more consecutive letters all of the same kind immediately preceded and succeeded by the other letter or the beginning or end of the row). The upper P per cent point $x(P)$ of R is the smallest x such that $\Pr\{R \geq x\} \leq P/100$, and the lower P per cent point $x'(P)$ of R is the largest x such that $\Pr\{R \leq x\} \leq P/100$. A dash indicates that there is no value with the required property. When n_1 and n_2 are large, R is approximately normally distributed with mean $\frac{2n_1n_2}{n_1+n_2}+1$ and variance $\frac{2n_1n_2(2n_1n_2-n_1-n_2)}{(n_1+n_2)^2(n_1+n_2-1)}$. Formulae for the calculation of this distribution are given by M. G. Kendall and A. Stuart, *The Advanced Theory of Statistics*, Vol. 2 (3rd edition, 1973), Griffin, London, Exercise 30.8.

UPPER PERCENTAGE POINTS

n_1	n_2 (P)	5	1	0·1
$n_1=3$	$n_2=4$	7	—	—
4	4	8	—	—
	5	9	9	—
	6	9	—	—
	7	9	—	—
5	5	9	10	—
	6	10	11	—
	7	10	11	—
	8	11	—	—
	9	11	—	—
5	10	11	—	—
6	6	11	12	—
	7	11	12	13
	8	12	13	—
	9	12	13	—
6	10	12	—	—
	11	13	—	—
	12	13	—	—
	13	13	—	—
	14	13	—	—
7	7	12	13	14
	8	13	14	15
	9	13	14	15
	10	13	15	—
	11	14	15	—
7	12	14	15	—
	13	14	—	—
	14	14	—	—
	15	15	—	—
	16	15	—	—
7	17	15	—	—
	18	15	—	—
	19	15	—	—
8	8	13	14	16
	9	14	15	16
8	10	14	15	17
	11	15	16	17
	12	15	16	—
	13	15	17	—
	14	16	17	—
8	15	16	17	—
	16	16	17	—

n_1	n_2 (P)	5	1	0·1
$n_1=8$	$n_2=17$	16	—	—
	18	16	—	—
	19	16	—	—
	20	17	—	—
9	9	14	16	17
9	10	15	16	18
	11	15	17	18
	12	16	17	19
	13	16	18	19
	14	17	18	19
9	15	17	18	—
	16	17	18	—
	17	17	19	—
	18	18	19	—
	19	18	19	—
9	20	18	19	—
10	10	16	17	18
	11	16	18	19
	12	17	18	20
	13	17	19	20
10	14	17	19	20
	15	18	19	21
	16	18	20	21
	17	18	20	21
	18	19	20	—
10	19	19	20	—
	20	19	20	—
11	11	17	18	20
	12	17	19	20
	13	18	19	21
11	14	18	20	21
	15	19	20	22
	16	19	21	22
	17	19	21	22
	18	20	21	23
11	19	20	22	23
	20	20	22	23
12	12	18	19	21
	13	18	20	22
	14	19	21	22
12	15	19	21	23

n_1	n_2 (P)	5	1	0·1
$n_1=12$	$n_2=16$	20	22	23
	17	20	22	24
	18	21	22	24
	19	21	23	24
	20	21	23	24
13	13	19	21	23
	14	20	21	23
	15	20	22	24
	16	21	22	24
	17	21	23	25
13	18	21	23	25
	19	22	24	25
	20	22	24	26
14	14	20	22	24
	15	21	23	24
14	16	21	23	25
	17	22	24	25
	18	22	24	26
	19	23	24	26
	20	23	25	27
15	15	21	23	25
	16	22	24	26
	17	22	24	26
	18	23	25	27
	19	23	25	27
15	20	24	26	28
16	16	23	24	26
	17	23	25	27
	18	24	26	28
	19	24	26	28
16	20	25	26	29
17	17	24	26	28
	18	24	26	28
	19	25	27	29
	20	25	27	29
18	18	25	27	29
	19	25	27	30
	20	26	28	30
19	19	26	28	30
	20	27	29	31
20	20	27	29	31

TABLE 18. PERCENTAGE POINTS OF THE DISTRIBUTION OF THE NUMBER OF RUNS

LOWER PERCENTAGE POINTS

n_1	n_2	P: 5	1	0·1
2	8	2	—	—
	9	2	—	—
	10	2	—	—
	11	2	—	—
	12	2	—	—
2	13	2	—	—
	14	2	—	—
	15	2	—	—
	16	2	—	—
	17	2	—	—
2	18	2	—	—
	19	2	2	—
	20	2	2	—
3	5	2	—	—
	6	2	—	—
3	7	2	—	—
	8	2	—	—
	9	2	2	—
	10	3	2	—
	11	3	2	—
3	12	3	2	—
	13	3	2	—
	14	3	2	—
	15	3	2	—
	16	3	2	—
3	17	3	2	—
	18	3	2	—
	19	3	2	—
	20	3	2	—
4	4	2	—	—
4	5	2	—	—
	6	3	2	—
	7	3	2	—
	8	3	2	—
	9	3	2	—
4	10	3	2	—
	11	3	2	—
	12	4	3	—
	13	4	3	2
	14	4	3	2
4	15	4	3	2
	16	4	3	2
	17	4	3	2
	18	4	3	2
	19	4	3	2
4	20	4	3	2
5	5	3	2	—
	6	3	2	—
	7	3	2	—
	8	3	2	—
5	9	4	3	2

n_1	n_2	P: 5	1	0·1
5	10	4	3	2
	11	4	3	2
	12	4	3	2
	13	4	3	2
	14	5	3	2
5	15	5	4	2
	16	5	4	2
	17	5	4	3
	18	5	4	3
	19	5	4	3
5	20	5	4	3
6	6	3	2	—
	7	4	3	—
	8	4	3	2
	9	4	3	2
6	10	5	3	2
	11	5	4	2
	12	5	4	3
	13	5	4	3
	14	5	4	3
6	15	6	4	3
	16	6	4	3
	17	6	5	3
	18	6	5	3
	19	6	5	3
6	20	6	5	4
7	7	4	3	2
	8	4	3	2
	9	5	4	2
	10	5	4	3
7	11	5	4	3
	12	6	4	3
	13	6	5	3
	14	6	5	3
	15	6	5	3
7	16	6	5	4
	17	7	5	4
	18	7	5	4
	19	7	6	4
	20	7	6	4
8	8	5	4	2
	9	5	4	3
	10	6	4	3
	11	6	5	3
	12	6	5	3
8	13	6	5	4
	14	7	5	4
	15	7	5	4
	16	7	6	4
	17	7	6	4
8	18	8	6	4

n_1	n_2	P: 5	1	0·1
8	19	8	6	5
	20	8	6	5
9	9	6	4	3
	10	6	5	3
	11	6	5	3
9	12	7	5	4
	13	7	6	4
	14	7	6	4
	15	8	6	4
	16	8	6	5
9	17	8	7	5
	18	8	7	5
	19	8	7	5
	20	9	7	5
10	10	6	5	4
10	11	7	5	4
	12	7	6	4
	13	8	6	4
	14	8	6	5
	15	8	7	5
10	16	8	7	5
	17	9	7	5
	18	9	7	6
	19	9	8	6
	20	9	8	6
11	11	7	6	4
	12	8	6	5
	13	8	6	5
	14	8	7	5
	15	9	7	5
11	16	9	7	6
	17	9	8	6
	18	10	8	6
	19	10	8	6
	20	10	8	7
12	12	8	7	5
	13	9	7	5
	14	9	7	5
	15	9	8	6
	16	10	8	6
12	17	10	8	6
	18	10	8	7
	19	10	9	7
	20	11	9	7
13	13	9	7	5
13	14	9	8	6
	15	10	8	6
	16	10	8	6
	17	10	9	7
	18	11	9	7
13	19	11	9	7

TABLE 18. PERCENTAGE POINTS OF THE DISTRIBUTION OF THE NUMBER OF RUNS

LOWER PERCENTAGE POINTS

n_1	n_2	P 5	1	0·1		n_1	n_2	P 5	1	0·1		n_1	n_2	P 5	1	0·1
13	20	11	10	8		15	17	11	10	8		17	18	13	11	9
14	14	10	8	6			18	12	10	8			19	13	11	9
	15	10	8	7			19	12	10	8			20	13	11	9
	16	11	9	7			20	12	11	8		18	18	13	11	9
	17	11	9	7		16	16	11	10	8			19	14	12	9
14	18	11	9	7		16	17	12	10	8		18	20	14	12	10
	19	12	10	8			18	12	10	8		19	19	14	12	10
	20	12	10	8			19	13	11	9			20	14	12	10
15	15	11	9	7			20	13	11	9		20	20	15	13	11
	16	11	9	7		17	17	12	10	8						

TABLE 19. UPPER PERCENTAGE POINTS OF THE TWO-SAMPLE KOLMOGOROV–SMIRNOV DISTRIBUTION

This table gives percentage points of

$$D(n_1, n_2) = \sup |F_1(x) - F_2(x)|,$$

where $F_1(x)$ and $F_2(x)$ are the empirical distribution functions of two independent random samples of sizes n_1 and n_2 respectively, $n_1 \leqslant n_2 \leqslant 20$ and $n_1 = n_2 \leqslant 100$, from the same population with a continuous distribution function; the function tabulated $d(P)$ is the smallest d such that Pr $\{n_1 n_2 D(n_1, n_2) \geqslant d\} \leqslant P/100$. A dash indicates that there is no value with the required property. A test of the hypothesis that two random samples of sizes n_1 and n_2 respectively have the same continuous distribution function is provided by rejecting at the P per cent level if $n_1 n_2 D(n_1, n_2) \geqslant d(P)$. When n_1 and n_2 are large, percentage points of $\sqrt{\dfrac{n_1 n_2}{n_1 + n_2}} D(n_1, n_2)$ are approximately given by those in Table 23 with $n = \infty$. Formulae for the calculation of this table are given by P. J. Kim and R. I. Jennrich, 'Tables of the exact sampling distribution of the two-sample Kolmogorov–Smirnov criterion D_{mn}, $m \leqslant n$', *Selected Tables in Mathematical Statistics*, Vol. 1 (1973), American Mathematical Society, Providence, R.I.

n_1	n_2	P 10	5	2·5	1	0·1		n_1	n_2	P 10	5	2·5	1	0·1
2	5	10	—	—	—	—		3	17	36	42	45	48	—
	6	12	—	—	—	—			18	39	45	48	51	—
	7	14	—	—	—	—			19	42	45	51	54	—
	8	16	16	—	—	—			20	42	48	51	57	—
	9	18	18	—	—	—		4	4	16	16	—	—	—
2	10	18	20	—	—	—		4	5	16	20	20	—	—
	11	20	22	—	—	—			6	18	20	24	24	—
	12	22	24	24	—	—			7	21	24	28	28	—
	13	24	26	26	—	—			8	24	28	28	32	—
	14	24	26	28	—	—			9	27	28	32	36	—
2	15	26	28	30	—	—		4	10	28	30	36	36	—
	16	28	30	32	—	—			11	29	33	36	40	—
	17	30	32	34	—	—			12	36	36	40	44	—
	18	32	34	36	—	—			13	35	39	44	48	52
	19	32	36	38	38	—			14	38	42	44	48	56
2	20	34	38	40	40	—		4	15	40	44	45	52	60
3	3	9	—	—	—	—			16	44	48	52	56	64
	4	12	—	—	—	—			17	44	48	52	60	68
	5	15	15	—	—	—			18	46	50	54	60	72
	6	15	18	18	—	—			19	49	53	57	64	76
3	7	18	21	21	—	—		4	20	52	60	64	68	76
	8	21	21	24	—	—		5	5	20	25	25	25	—
	9	21	24	27	27	—			6	24	24	30	30	—
	10	24	27	30	30	—			7	25	28	30	35	—
	11	27	30	30	33	—			8	27	30	32	35	—
3	12	27	30	33	36	—		5	9	30	35	36	40	45
	13	30	33	36	39	—			10	35	40	40	45	50
	14	33	36	39	42	—			11	35	39	44	45	55
	15	33	36	39	42	—			12	36	43	45	50	60
	16	36	39	42	45	—			13	40	45	47	52	65

TABLE 19. UPPER PERCENTAGE POINTS OF THE TWO-SAMPLE KOLMOGOROV–SMIRNOV DISTRIBUTION

n_1	n_2	P	10	5	2·5	1	0·1
5	14		42	46	51	56	70
	15		50	55	55	60	70
	16		48	54	59	64	75
	17		50	55	60	68	80
	18		52	60	65	70	85
5	19		56	61	66	71	85
	20		60	65	75	80	90
6	6		30	30	36	36	—
	7		28	30	35	36	—
	8		30	34	36	40	48
6	9		33	39	42	45	54
	10		36	40	44	48	60
	11		38	43	48	54	66
	12		48	48	54	60	66
	13		46	52	54	60	72
6	14		48	54	58	64	78
	15		51	57	63	69	84
	16		54	60	64	72	84
	17		56	62	67	73	85
	18		66	72	78	84	96
6	19		64	70	76	83	96
	20		66	72	78	88	100
7	7		35	42	42	42	49
	8		34	40	41	48	56
	9		36	42	45	49	63
7	10		40	46	49	53	63
	11		44	48	52	59	70
	12		46	53	56	60	72
	13		50	56	58	65	78
	14		56	63	70	77	84
7	15		56	62	68	75	90
	16		59	64	73	77	96
	17		61	68	77	84	98
	18		65	72	80	87	101
	19		69	76	84	91	107
7	20		72	79	86	93	112
8	8		40	48	48	56	64
	9		40	46	48	55	64
	10		44	48	54	60	70
	11		48	53	58	64	77
8	12		52	60	64	68	80
	13		54	62	65	72	88
	14		58	64	70	76	90
	15		60	67	74	81	97
	16		72	80	80	88	104
8	17		68	77	80	88	111
	18		72	80	86	94	112
	19		74	82	90	98	117
	20		80	88	96	104	124
9	9		54	54	63	63	72
9	10		50	53	60	63	80
	11		52	59	63	70	81
	12		57	63	69	75	87

n_1	n_2	P	10	5	2·5	1	0·1
9	13		59	65	72	78	91
	14		63	70	76	84	98
	15		69	75	81	90	105
	16		69	78	85	94	110
	17		74	82	90	99	117
9	18		81	90	99	108	126
	19		80	89	98	107	126
	20		84	93	100	111	133
10	10		60	70	70	80	90
	11		57	60	68	77	89
10	12		60	66	72	80	96
	13		64	70	77	84	100
	14		68	74	82	90	106
	15		75	80	90	100	115
	16		76	84	90	100	118
10	17		79	89	96	106	126
	18		82	92	100	108	132
	19		85	94	103	113	133
	20		100	110	120	130	150
11	11		66	77	77	88	99
11	12		64	72	76	86	99
	13		67	75	84	91	108
	14		73	82	87	96	115
	15		76	84	94	102	120
	16		80	89	96	106	127
11	17		85	93	102	110	132
	18		88	97	107	118	140
	19		92	102	111	122	146
	20		96	107	116	127	154
12	12		72	84	96	96	120
12	13		71	81	84	95	117
	14		78	86	94	104	120
	15		84	93	99	108	129
	16		88	96	104	116	136
	17		90	100	108	119	141
12	18		96	108	120	126	150
	19		99	108	120	130	156
	20		104	116	124	140	164
13	13		91	91	104	117	130
	14		78	89	100	104	129
13	15		87	96	104	115	137
	16		91	101	111	121	143
	17		96	105	114	127	152
	18		99	110	120	131	156
	19		104	114	126	138	164
13	20		108	120	130	143	169
14	14		98	112	112	126	154
	15		92	98	110	123	140
	16		96	106	116	126	152
	17		100	111	122	134	159
14	18		104	116	126	140	166
	19		110	121	133	148	176
	20		114	126	138	152	180

TABLE 19. UPPER PERCENTAGE POINTS OF THE TWO-SAMPLE KOLMOGOROV–SMIRNOV DISTRIBUTION

n_1	n_2	P	10	5	2·5	1	0·1
15	15		105	120	135	135	165
	16		101	114	119	133	162
	17		105	116	129	142	165
	18		111	123	135	147	174
	19		114	127	141	152	180
15	20		125	135	150	160	195
16	16		112	128	144	160	176
	17		109	124	136	143	174
	18		116	128	140	154	186
	19		120	133	145	160	190

n_1	n_2	P	10	5	2·5	1	0·1
16	20		128	140	156	168	200
17	17		136	136	153	170	204
	18		118	133	148	164	187
	19		126	141	151	166	200
	20		130	146	160	175	209
18	18		144	162	162	180	216
	19		133	142	159	176	212
	20		136	152	166	182	214
19	19		152	171	190	190	228
	20		144	160	169	187	225

$n_1 = n_2$	10	5	2·5	1	0·1
20	160	180	200	220	260
21	168	189	210	231	273
22	198	198	220	242	286
23	207	230	230	253	299
24	216	240	264	288	336
25	225	250	275	300	350
26	234	260	286	312	364
27	243	270	297	324	405
28	280	308	336	364	420
29	290	319	348	377	435
30	300	330	360	390	450
31	310	341	372	403	496
32	320	352	384	416	512
33	330	396	396	462	528
34	374	408	442	476	544
35	385	420	455	490	595
36	396	432	468	504	612
37	407	444	481	518	629
38	418	456	494	570	646
39	429	468	546	585	702
40	440	520	560	600	720
41	492	533	574	615	738
42	504	546	588	630	756
43	516	559	602	688	774
44	528	572	616	704	836
45	540	585	675	720	855
46	552	644	690	736	874
47	564	658	705	752	893
48	576	672	720	768	912
49	637	686	735	833	980
50	650	700	750	850	1000
51	663	714	765	867	1020
52	676	728	832	884	1040
53	689	742	848	901	1060
54	702	810	864	918	1134
55	715	825	880	990	1155
56	728	840	896	1008	1176
57	798	855	912	1026	1197
58	812	870	928	1044	1218
59	826	885	1003	1062	1298
60	840	900	1020	1080	1320

$n_1 = n_2$	10	5	2·5	1	0·1
60	840	900	1020	1080	1320
61	854	915	1037	1098	1342
62	868	992	1054	1178	1364
63	882	1008	1071	1197	1386
64	896	1024	1088	1216	1408
65	910	1040	1105	1235	1495
66	990	1056	1122	1254	1518
67	1005	1072	1206	1273	1541
68	1020	1088	1224	1292	1564
69	1035	1104	1242	1380	1587
70	1050	1190	1260	1400	1610
71	1065	1207	1278	1420	1704
72	1080	1224	1296	1440	1728
73	1095	1241	1314	1460	1752
74	1110	1258	1332	1480	1776
75	1125	1275	1425	1500	1800
76	1216	1292	1444	1596	1824
77	1232	1309	1463	1617	1925
78	1248	1326	1482	1638	1950
79	1264	1422	1501	1659	1975
80	1280	1440	1520	1680	2000
81	1296	1458	1539	1701	2025
82	1312	1476	1558	1722	2050
83	1328	1494	1660	1743	2075
84	1344	1512	1680	1848	2184
85	1360	1530	1700	1870	2210
86	1462	1548	1720	1892	2236
87	1479	1566	1740	1914	2262
88	1496	1672	1760	1936	2288
89	1513	1691	1780	1958	2314
90	1530	1710	1800	1980	2430
91	1547	1729	1820	2002	2457
92	1564	1748	1932	2116	2484
93	1581	1767	1953	2139	2511
94	1598	1786	1974	2162	2538
95	1615	1805	1995	2185	2565
96	1632	1824	2016	2208	2592
97	1746	1843	2037	2231	2716
98	1764	1960	2058	2254	2744
99	1782	1980	2079	2277	2772
100	1800	2000	2100	2300	2800

TABLE 20. PERCENTAGE POINTS OF WILCOXON'S SIGNED-RANK DISTRIBUTION

This table gives lower percentage points of W^+, the sum of the ranks of the positive observations in a ranking in order of increasing absolute magnitude of a random sample of size n from a continuous distribution, symmetric about zero. The function tabulated $x(P)$ is the largest x such that $\Pr\{W^+ \leq x\} \leq P/100$. A dash indicates that there is no value with the required property. W^-, the sum of the ranks of the negative observations, has the same distribution as W^+, with mean $\frac{1}{4}n(n+1)$ and variance $\frac{1}{12}n(n+1)(n+\frac{1}{2})$. A test of the hypothesis that a random sample of size n has arisen from a continuous distribution symmetric about $\mu = 0$ against the alternative that $\mu < 0$ is provided by rejecting at the P per cent level if $W^+ \leq x(P)$; a similar test against $\mu > 0$ is provided by rejecting at the P per cent level if $W^- \leq x(P)$, and, against $\mu \neq 0$, one rejects at the $2P$ per cent level if W, the smaller of W^+ and W^-, is less than or equal to $x(P)$. When $n > 85$, W^+ is approximately normally distributed.

Formulae for the calculation of this table are given by F. Wilcoxon, S. K. Katti and R. A. Wilcox, 'Critical values and probability levels for the Wilcoxon rank sum test and the Wilcoxon signed rank test', *Selected Tables in Mathematical Statistics*, Vol. 1 (1973), American Mathematical Society, Providence, R.I.

P	5	2·5	1	0·5	0·1
$n = 5$	0	—	—	—	—
6	2	0	—	—	—
7	3	2	0	—	—
8	5	3	1	0	—
9	8	5	3	1	—
10	10	8	5	3	0
11	13	10	7	5	1
12	17	13	9	7	2
13	21	17	12	9	4
14	25	21	15	12	6
15	30	25	19	15	8
16	35	29	23	19	11
17	41	34	27	23	14
18	47	40	32	27	18
19	53	46	37	32	21
20	60	52	43	37	26
21	67	58	49	42	30
22	75	65	55	48	35
23	83	73	62	54	40
24	91	81	69	61	45
25	100	89	76	68	51
26	110	98	84	75	58
27	119	107	92	83	64
28	130	116	101	91	71
29	140	126	110	100	79
30	151	137	120	109	86
31	163	147	130	118	94
32	175	159	140	128	103
33	187	170	151	138	112
34	200	182	162	148	121
35	213	195	173	159	131
36	227	208	185	171	141
37	241	221	198	182	151
38	256	235	211	194	162
39	271	249	224	207	173
40	286	264	238	220	185
41	302	279	252	233	197
42	319	294	266	247	209
43	336	310	281	261	222
44	353	327	296	276	235
45	371	343	312	291	249

P	5	2·5	1	0·5	0·1
$n = 45$	371	343	312	291	249
46	389	361	328	307	263
47	407	378	345	322	277
48	426	396	362	339	292
49	446	415	379	355	307
50	466	434	397	373	323
51	486	453	416	390	339
52	507	473	434	408	355
53	529	494	454	427	372
54	550	514	473	445	389
55	573	536	493	465	407
56	595	557	514	484	425
57	618	579	535	504	443
58	642	602	556	525	462
59	666	625	578	546	482
60	690	648	600	567	501
61	715	672	623	589	521
62	741	697	646	611	542
63	767	721	669	634	563
64	793	747	693	657	584
65	820	772	718	681	606
66	847	798	742	705	628
67	875	825	768	729	651
68	903	852	793	754	674
69	931	879	819	779	697
70	960	907	846	805	721
71	990	936	873	831	745
72	1020	964	901	858	770
73	1050	994	928	884	795
74	1081	1023	957	912	821
75	1112	1053	986	940	847
76	1144	1084	1015	968	873
77	1176	1115	1044	997	900
78	1209	1147	1075	1026	927
79	1242	1179	1105	1056	955
80	1276	1211	1136	1086	983
81	1310	1244	1168	1116	1011
82	1345	1277	1200	1147	1040
83	1380	1311	1232	1178	1070
84	1415	1345	1265	1210	1099
85	1451	1380	1298	1242	1130

TABLE 21. PERCENTAGE POINTS OF THE MANN–WHITNEY DISTRIBUTION

Consider two independent random samples of sizes n_1 and n_2 respectively ($n_1 \leqslant n_2$) from two continuous populations, A and B. Let all n_1+n_2 observations be ranked in increasing order and let R_A and R_B denote the sums of the ranks of the observations in samples A and B respectively. This table gives lower percentage points of $U_A = R_A - \frac{1}{2}n_1(n_1+1)$; the function tabulated $x(P)$ is the largest x such that, on the assumption that populations A and B are identical, $\Pr\{U_A \leqslant x\} \leqslant P/100$. A dash indicates that there is no value with the required property. On the same assumption, $U_B = R_B - \frac{1}{2}n_2(n_2+1)$ has the same distribution as U_A, with mean $\frac{1}{2}n_1 n_2$ and variance $\frac{1}{12}n_1 n_2(n_1+n_2+1)$. A test of the hypothesis that the two populations are identical, and in particular that their respective means μ_A, μ_B are equal, against the alternative $\mu_A > \mu_B$ is provided by rejecting at level P per cent if $U_B \leqslant x(P)$, and a similar test against $\mu_A < \mu_B$ is provided by rejecting at the P per cent level if $U_A \leqslant x(P)$. For a test against both alternatives one rejects at the $2P$ per cent level if U, the smaller of U_A and U_B, is less than or equal to $x(P)$. If n_1 and n_2 are large U_A is approximately normally distributed. Note also that $U_A+U_B = n_1 n_2$.

Formulae for the calculation of this distribution (which is also referred to as the Wilcoxon rank–sum or Wilcoxon/Mann–Whitney distribution) are given by F. Wilcoxon, S. K. Katti and R. A. Wilcox, 'Critical values and probability levels for the Wilcoxon rank sum test and the Wilcoxon signed rank test', *Selected Tables in Mathematical Statistics*, Vol. 1 (1973), American Mathematical Society, Providence, R.I.

n_1	n_2	P	5	2·5	1	0·5	0·1
2	5		0	—	—	—	—
	6		0	—	—	—	—
	7		0	—	—	—	—
	8		1	0	—	—	—
	9		1	0	—	—	—
2	10		1	0	—	—	—
	11		1	0	—	—	—
	12		2	1	—	—	—
	13		2	1	0	—	—
	14		3	1	0	—	—
2	15		3	1	0	—	—
	16		3	1	0	—	—
	17		3	2	0	—	—
	18		4	2	0	—	—
	19		4	2	1	0	—
2	20		4	2	1	0	—
3	3		0	—	—	—	—
	4		0	—	—	—	—
	5		1	0	—	—	—
	6		2	1	—	—	—
3	7		2	1	0	—	—
	8		3	2	0	—	—
	9		4	2	1	0	—
	10		4	3	1	0	—
	11		5	3	1	0	—
3	12		5	4	2	1	—
	13		6	4	2	1	—
	14		7	5	2	1	—
	15		7	5	3	2	—
	16		8	6	3	2	—
3	17		9	6	4	2	0
	18		9	7	4	2	0
	19		10	7	4	3	0
	20		11	8	5	3	0
4	4		1	0	—	—	—
4	5		2	1	0	—	—
	6		3	2	1	0	—
	7		4	3	1	0	—
	8		5	4	2	1	—
	9		6	4	3	1	—

n_1	n_2	P	5	2·5	1	0·5	0·1
4	10		7	5	3	2	0
	11		8	6	4	2	0
	12		9	7	5	3	0
	13		10	8	5	3	1
	14		11	9	6	4	1
	15		12	10	7	5	1
	16		14	11	7	5	2
	17		15	11	8	6	2
	18		16	12	9	6	3
	19		17	13	9	7	3
4	20		18	14	10	8	3
5	5		4	2	1	0	—
	6		5	3	2	1	—
	7		6	5	3	1	—
	8		8	6	4	2	0
5	9		9	7	5	3	1
	10		11	8	6	4	1
	11		12	9	7	5	2
	12		13	11	8	6	2
	13		15	12	9	7	3
5	14		16	13	10	7	3
	15		18	14	11	8	4
	16		19	15	12	9	5
	17		20	17	13	10	5
	18		22	18	14	11	6
5	19		23	19	15	12	7
	20		25	20	16	13	7
6	6		7	5	3	2	—
	7		8	6	4	3	0
	8		10	8	6	4	1
6	9		12	10	7	5	2
	10		14	11	8	6	3
	11		16	13	9	7	4
	12		17	14	11	9	4
	13		19	16	12	10	5
6	14		21	17	13	11	6
	15		23	19	15	12	7
	16		25	21	16	13	8
	17		26	22	18	15	9
	18		28	24	19	16	10

TABLE 21. PERCENTAGE POINTS OF THE MANN–WHITNEY DISTRIBUTION

n_1	n_2	P = 5	2.5	1	0.5	0.1
6	19	30	25	20	17	11
	20	32	27	22	18	12
7	7	11	8	6	4	1
	8	13	10	7	6	2
	9	15	12	9	7	3
7	10	17	14	11	9	5
	11	19	16	12	10	6
	12	21	18	14	12	7
	13	24	20	16	13	8
	14	26	22	17	15	9
7	15	28	24	19	16	10
	16	30	26	21	18	11
	17	33	28	23	19	13
	18	35	30	24	21	14
	19	37	32	26	22	15
7	20	39	34	28	24	16
8	8	15	13	9	7	4
	9	18	15	11	9	5
	10	20	17	13	11	6
	11	23	19	15	13	8
8	12	26	22	17	15	9
	13	28	24	20	17	11
	14	31	26	22	18	12
	15	33	29	24	20	14
	16	36	31	26	22	15
8	17	39	34	28	24	17
	18	41	36	30	26	18
	19	44	38	32	28	20
	20	47	41	34	30	21
9	9	21	17	14	11	7
9	10	24	20	16	13	8
	11	27	23	18	16	10
	12	30	26	21	18	12
	13	33	28	23	20	14
	14	36	31	26	22	15
9	15	39	34	28	24	17
	16	42	37	31	27	19
	17	45	39	33	29	21
	18	48	42	36	31	23
	19	51	45	38	33	25
9	20	54	48	40	36	26
10	10	27	23	19	16	10
	11	31	26	22	18	12
	12	34	29	24	21	14
	13	37	33	27	24	17
10	14	41	36	30	26	19
	15	44	39	33	29	21
	16	48	42	36	31	23
	17	51	45	38	34	25
	18	55	48	41	37	27
10	19	58	52	44	39	29
	20	62	55	47	42	32
11	11	34	30	25	21	15

n_1	n_2	P = 5	2.5	1	0.5	0.1
11	12	38	33	28	24	17
	13	42	37	31	27	20
	14	46	40	34	30	22
	15	50	44	37	33	24
	16	54	47	41	36	27
11	17	57	51	44	39	29
	18	61	55	47	42	32
	19	65	58	50	45	34
	20	69	62	53	48	37
12	12	42	37	31	27	20
12	13	47	41	35	31	23
	14	51	45	38	34	25
	15	55	49	42	37	28
	16	60	53	46	41	31
	17	64	57	49	44	34
12	18	68	61	53	47	37
	19	72	65	56	51	40
	20	77	69	60	54	42
13	13	51	45	39	34	26
	14	56	50	43	38	29
13	15	61	54	47	42	32
	16	65	59	51	45	35
	17	70	63	55	49	38
	18	75	67	59	53	42
	19	80	72	63	57	45
13	20	84	76	67	60	48
14	14	61	55	47	42	32
	15	66	59	51	46	36
	16	71	64	56	50	39
	17	77	69	60	54	43
14	18	82	74	65	58	46
	19	87	78	69	63	50
	20	92	83	73	67	54
15	15	72	64	56	51	40
	16	77	70	61	55	43
15	17	83	75	66	60	47
	18	88	80	70	64	51
	19	94	85	75	69	55
	20	100	90	80	73	59
16	16	83	75	66	60	48
16	17	89	81	71	65	52
	18	95	86	76	70	56
	19	101	92	82	74	60
	20	107	98	87	79	65
17	17	96	87	77	70	57
17	18	102	93	82	75	61
	19	109	99	88	81	66
	20	115	105	93	86	70
18	18	109	99	88	81	66
	19	116	106	94	87	71
18	20	123	112	100	92	76
19	19	123	113	101	93	77
	20	130	119	107	99	82
20	20	138	127	114	105	88

TABLE 22A. EXPECTED VALUES OF NORMAL ORDER STATISTICS (NORMAL SCORES)

Suppose that n independent observations, normally distributed with zero mean and unit variance, are arranged in decreasing order, and let the rth value in this ordering be denoted by $Z(r)$. This table gives expected values $E(n, r)$ of $Z(r)$ for $r \leqslant \frac{1}{2}(n+1)$; when $r > \frac{1}{2}(n+1)$ use

$$E(n, r) = -E(n, n+1-r).$$

The values $E(n, r)$ are often referred to as *normal scores*; they have a number of applications in statistics. In carrying out calculations for some of these applications the sums of squares of normal scores are often required: they are provided in Table 22 B.

$n =$	1	2	3	4	5	6	7	8	9	10
$r = 1$	0·0000	0·5642	0·8463	1·0294	1·1630	1·2672	1·3522	1·4236	1·4850	1·5388
2			·0000	0·2970	0·4950	0·6418	0·7574	0·8522	0·9323	1·0014
3					0·0000	0·2015	0·3527	0·4728	0·5720	0·6561
4							0·0000	0·1525	0·2745	0·3758
5									0·0000	0·1227

$n =$	11	12	13	14	15	16	17	18	19	20
$r = 1$	1·5864	1·6292	1·6680	1·7034	1·7359	1·7660	1·7939	1·8200	1·8445	1·8675
2	1·0619	1·1157	1·1641	1·2079	1·2479	1·2847	1·3188	1·3504	1·3799	1·4076
3	0·7288	0·7928	0·8498	0·9011	0·9477	0·9903	1·0295	1·0657	1·0995	1·1309
4	0·4620	0·5368	0·6029	0·6618	0·7149	0·7632	0·8074	0·8481	0·8859	0·9210
5	0·2249	0·3122	0·3883	0·4556	0·5157	0·5700	0·6195	0·6648	0·7066	0·7454
6	·0000	·1026	·1905	·2673	·3353	·3962	·4513	·5016	·5477	·5903
7			·0000	·0882	·1653	·2338	·2952	·3508	·4016	·4483
8					·0000	·0773	·1460	·2077	·2637	·3149
9							·0000	·0688	·1307	·1870
10									0·0000	0·0620

$n =$	21	22	23	24	25	26	27	28	29	30
$r = 1$	1·8892	1·9097	1·9292	1·9477	1·9653	1·9822	1·9983	2·0137	2·0285	2·0428
2	1·4336	1·4582	·4814	·5034	·5243	·5442	·5633	1·5815	1·5989	1·6156
3	1·1605	1·1882	·2144	·2392	·2628	·2851	·3064	1·3267	1·3462	1·3648
4	0·9538	0·9846	·0136	·0409	·0668	·0914	·1147	1·1370	1·1582	1·1786
5	0·7815	0·8153	0·8470	0·8768	0·9050	0·9317	0·9570	0·9812	1·0041	1·0261
6	·6298	·6667	·7012	·7335	·7641	·7929	·8202	·8462	0·8708	0·8944
7	·4915	·5316	·5690	·6040	·6369	·6679	·6973	·7251	0·7515	0·7767
8	·3620	·4056	·4461	·4839	·5193	·5527	·5841	·6138	0·6420	0·6689
9	·2384	·2858	·3297	·3705	·4086	·4444	·4780	·5098	0·5398	0·5683
10	0·1184	0·1700	0·2175	0·2616	0·3027	0·3410	0·3771	0·4110	0·4430	0·4733
11	·0000	·0564	·1081	·1558	·2001	·2413	·2798	·3160	·3501	·3824
12			·0000	·0518	·0995	·1439	·1852	·2239	·2602	·2945
13					·0000	·0478	·0922	·1336	·1724	·2088
14							·0000	·0444	·0859	·1247
15									0·0000	0·0415

TABLE 22A. EXPECTED VALUES OF NORMAL ORDER STATISTICS (NORMAL SCORES)

$n =$	31	32	33	34	35	36	37	38	39	40
$r = 1$	2·0565	2·0697	2·0824	2·0947	2·1066	2·1181	2·1293	2·1401	2·1506	2·1608
2	1·6317	1·6471	1·6620	1·6764	1·6902	1·7036	1·7166	1·7291	1·7413	1·7531
3	1·3827	1·3998	1·4164	1·4323	1·4476	1·4624	1·4768	1·4906	1·5040	1·5170
4	1·1980	1·2167	1·2347	1·2520	1·2686	1·2847	1·3002	1·3151	1·3296	1·3437
5	1·0471	1·0672	1·0865	1·1051	1·1230	1·1402	1·1568	1·1728	1·1883	1·2033
6	0·9169	0·9384	0·9590	0·9789	0·9979	1·0162	1·0339	1·0509	1·0674	1·0833
7	0·8007	0·8236	0·8455	0·8666	0·8868	0·9063	0·9250	0·9430	0·9604	0·9772
8	0·6944	0·7187	0·7420	0·7643	0·7857	0·8063	0·8261	0·8451	0·8634	0·8811
9	0·5955	0·6213	0·6460	0·6695	0·6921	0·7138	0·7346	0·7547	0·7740	0·7926
10	0·5021	0·5294	0·5555	0·5804	0·6043	0·6271	0·6490	0·6701	0·6904	0·7099
11	·4129	·4418	·4694	·4957	·5208	·5449	·5679	·5900	·6113	·6318
12	·3269	·3575	·3867	·4144	·4409	·4662	·4904	·5136	·5359	·5574
13	·2432	·2757	·3065	·3358	·3637	·3903	·4158	·4401	·4635	·4859
14	·1613	·1957	·2283	·2592	·2886	·3166	·3434	·3689	·3934	·4169
15	0·0804	0·1169	0·1515	0·1841	0·2151	0·2446	0·2727	0·2995	0·3252	0·3498
16	·0000	·0389	·0755	·1101	·1428	·1739	·2034	·2316	·2585	·2842
17			·0000	·0366	·0712	·1040	·1351	·1647	·1929	·2199
18					·0000	·0346	·0674	·0985	·1282	·1564
19							·0000	·0328	·0640	·0936
20									0·0000	0·0312

$n =$	41	42	43	44	45	46	47	48	49	50
$r = 1$	2·1707	2·1803	2·1897	2·1988	2·2077	2·2164	2·2249	2·2331	2·2412	2·2491
2	1·7646	1·7757	1·7865	1·7971	1·8073	1·8173	1·8271	1·8366	1·8458	1·8549
3	1·5296	1·5419	1·5538	1·5653	1·5766	1·5875	1·5982	1·6086	1·6187	1·6286
4	1·3573	1·3705	1·3833	1·3957	1·4078	1·4196	1·4311	1·4422	1·4531	1·4637
5	1·2178	1·2319	1·2456	1·2588	1·2717	1·2842	1·2964	1·3083	1·3198	1·3311
6	1·0987	1·1136	1·1281	1·1421	1·1558	1·1690	1·1819	1·1944	1·2066	1·2185
7	0·9935	1·0092	1·0245	1·0392	1·0536	1·0675	1·0810	1·0942	1·1070	1·1195
8	0·8982	0·9148	0·9308	0·9463	0·9614	0·9760	0·9902	1·0040	1·0174	1·0304
9	0·8106	0·8279	0·8447	0·8610	0·8767	0·8920	0·9068	0·9213	0·9353	0·9489
10	0·7287	0·7469	0·7645	0·7815	0·7979	0·8139	0·8294	0·8444	0·8590	0·8732
11	·6515	·6705	·6889	·7067	·7238	·7405	·7566	·7723	·7875	·8023
12	·5780	·5979	·6171	·6356	·6535	·6709	·6877	·7040	·7198	·7351
13	·5075	·5283	·5483	·5676	·5863	·6044	·6219	·6388	·6552	·6712
14	·4394	·4611	·4820	·5022	·5217	·5405	·5586	·5763	·5933	·6099
15	0·3734	0·3960	0·4178	0·4389	0·4591	0·4787	0·4976	0·5159	0·5336	0·5508
16	·3089	·3326	·3553	·3772	·3983	·4187	·4383	·4573	·4757	·4935
17	·2457	·2704	·2942	·3170	·3390	·3602	·3806	·4003	·4194	·4379
18	·1835	·2093	·2341	·2579	·2808	·3029	·3241	·3446	·3644	·3836
19	·1219	·1490	·1749	·1997	·2236	·2465	·2686	·2899	·3105	·3304
20	0·0608	0·0892	0·1163	0·1422	0·1671	0·1910	0·2140	0·2361	0·2575	0·2781
21	·0000	·0297	·0580	·0851	·1111	·1360	·1599	·1830	·2051	·2265
22			·0000	·0283	·0555	·0814	·1064	·1303	·1534	·1756
23					·0000	·0271	·0531	·0781	·1020	·1251
24							·0000	·0260	·0509	·0749
25									0·0000	0·0250

TABLE 22B. SUMS OF SQUARES OF NORMAL SCORES

This table gives values of $S(n) = \sum_{r=1}^{n} [E(n, r)]^2$.

n	$S(n)$	n	$S(n)$	n	$S(n)$	n	$S(n)$	n	$S(n)$
		10	7.914	20	17.678	30	27.558	40	37.479
1	0.0000	11	8.879	21	18.663	31	28.549	41	38.473
2	0.6366	12	9.848	22	19.649	32	29.540	42	39.466
3	1.432	13	10.820	23	20.635	33	30.531	43	40.460
4	2.296	14	11.795	24	21.623	34	31.523	44	41.454
5	3.195	15	12.771	25	22.610	35	32.515	45	42.448
6	4.117	16	13.750	26	23.599	36	33.507	46	43.443
7	5.053	17	14.730	27	24.588	37	34.500	47	44.437
8	5.999	18	15.711	28	25.577	38	35.493	48	45.432
9	6.954	19	16.694	29	26.567	39	36.486	49	46.427
10	7.914	20	17.678	30	27.558	40	37.479	50	47.422

TABLE 23. UPPER PERCENTAGE POINTS OF THE ONE-SAMPLE KOLMOGOROV–SMIRNOV DISTRIBUTION

If $F_n(x)$ is the empirical distribution function of a random sample of size n from a population with continuous distribution function $F(x)$, the table gives percentage points of $D(n) = \sup |F_n(x) - F(x)|$; the function tabulated is $d(P)$ such that the probability that $n^{\frac{1}{2}}D(n)$ exceeds $d(P)$ is $P/100$. A test of the hypothesis that the sample has arisen from $F(x)$ is provided by rejecting at the P per cent level if $n^{\frac{1}{2}}D(n) \geqslant d(P)$. The distribution of $n^{\frac{1}{2}}D(n)$ tends to a limit as n tends to infinity and the percentage points of this distribution are given under $n = \infty$. This table was calculated using formulae given by J. Durbin, *Distribution Theory for Tests Based on the Sample Distribution Function* (1973), Society for Industrial and Applied Mathematics, Philadelphia, Pa., Section 2.4.

P	10	5	2.5	1	0.1
$n = 1$	0.950	0.975	0.9875	0.995	0.9995
2	1.098	1.191	1.256	1.314	1.383
3	1.102	1.226	1.330	1.436	1.595
4	1.130	1.248	1.348	1.468	1.701
5	1.139	1.260	1.370	1.495	1.747
6	.146	.272	.382	.510	.775
7	.154	.279	.391	.523	.797
8	.159	.285	.399	.532	.813
9	.162	.290	.404	.540	.825
10	1.166	1.294	1.409	1.546	1.835
11	.169	.298	.413	.551	.844
12	.171	.301	.417	.556	.851
13	.174	.303	.420	.559	.856
14	.176	.305	.423	.563	.862
15	1.177	1.308	1.425	1.565	1.866
16	.179	.309	.427	.568	.870
17	.180	.311	.429	.570	.874
18	.182	.313	.431	.572	.877
19	.183	.314	.432	.574	.880
20	1.184	1.315	1.434	1.576	1.882

P	10	5	2.5	1	0.1
$n = 20$	1.184	1.315	1.434	1.576	1.882
21	.185	.316	.435	.578	.884
22	.186	.317	.436	.579	.887
23	.187	.318	.438	.580	.889
24	.188	.319	.439	.582	.890
25	1.188	1.320	1.440	1.583	1.892
26	.189	.321	.440	.584	.894
27	.190	.322	.441	.585	.895
28	.190	.323	.442	.586	.897
29	.191	.323	.443	.587	.898
30	1.192	1.324	1.444	1.588	1.899
40	.196	.329	.449	.594	.908
50	.199	.332	.453	.598	.914
60	.201	.335	.456	.601	.918
70	.203	.337	.458	.604	.921
80	1.205	1.338	1.459	1.605	1.923
90	.206	.339	.461	.607	.925
100	.207	.340	.462	.608	.927
200	.212	.346	.467	.614	.935
∞	.224	.358	.480	.628	.949

TABLE 24. UPPER PERCENTAGE POINTS OF FRIEDMAN'S DISTRIBUTION

Consider nk observations, one for each combination of n blocks and k treatments, and set out the observations in an $n \times k$ table, the columns relating to treatments and the rows to blocks. Let the observations in each row be ranked from 1 to k, and let R_j ($j = 1, 2, ..., k$) denote the sum of the ranks in the jth column. This table gives percentage points of Friedman's statistic

$$M = \frac{12}{nk(k+1)} \sum_{j=1}^{k} R_j^2 - 3n(k+1)$$

on the assumption of no difference between the treatments; the function tabulated $x(P)$ is the smallest value x such that, on this assumption, $\Pr\{M \geqslant x\} \leqslant P/100$. A dash indicates that there is no value with the required property. A test of the hypothesis of no difference between the treatments is provided by rejecting at the P per cent level if $M \geqslant x(P)$. The limiting distribution of M as n tends to infinity is the χ^2-distribution with $k-1$ degrees of freedom (see Table 8) and the percentage points are given under $n = \infty$.

$k = 3$

P	10	5	2.5	1	0.1
$n = 3$	6.000	6.000	—	—	—
4	6.000	6.500	8.000	8.000	—
5	5.200	6.400	7.600	8.400	10.00
6	5.333	7.000	8.333	9.000	12.00
7	5.429	7.143	7.714	8.857	12.29
8	5.250	6.250	7.750	9.000	12.25
9	5.556	6.222	8.000	9.556	12.67
10	5.000	6.200	7.800	9.600	12.60
11	5.091	6.545	7.818	9.455	13.27
12	5.167	6.500	8.000	9.500	12.67
13	4.769	6.615	7.538	9.385	12.46
14	5.143	6.143	7.429	9.143	13.29
15	4.933	6.400	7.600	8.933	12.93
16	4.875	6.500	7.625	9.375	13.50
17	5.059	6.118	7.412	9.294	13.06
18	4.778	6.333	7.444	9.000	13.00
19	5.053	6.421	7.684	9.579	13.37
20	4.900	6.300	7.500	9.300	13.30
21	4.952	6.095	7.524	9.238	13.24
22	4.727	6.091	7.364	9.091	13.45
23	4.957	6.348	7.913	9.391	13.13
24	5.083	6.250	7.750	9.250	13.08
25	4.880	6.080	7.440	8.960	13.52
26	4.846	6.077	7.462	9.308	13.23
27	4.741	6.000	7.407	9.407	13.41
28	4.571	6.500	7.714	9.214	13.50
29	5.034	6.276	7.517	9.172	13.52
30	4.867	6.200	7.400	9.267	13.40
31	4.839	6.000	7.548	9.290	13.42
32	4.750	6.063	7.563	9.250	13.69
33	4.788	6.061	7.515	9.152	13.52
34	4.765	6.059	7.471	9.176	13.41
∞	4.605	5.991	7.378	9.210	13.82

$k = 4$

P	10	5	2.5	1	0.1
$n = 3$	6.600	7.400	8.200	9.000	—
4	6.300	7.800	8.400	9.600	11.10
5	6.360	7.800	8.760	9.960	12.60
6	6.400	7.600	8.800	10.20	12.80
7	6.429	7.800	9.000	10.54	13.46
8	6.300	7.650	9.000	10.50	13.80
9	6.200	7.667	8.867	10.73	14.07
10	6.360	7.680	9.000	10.68	14.52
11	6.273	7.691	9.000	10.75	14.56
12	6.300	7.700	9.100	10.80	14.80
13	6.138	7.800	9.092	10.85	14.91
14	6.343	7.714	9.086	10.89	15.09
15	6.280	7.720	9.160	10.92	15.08
16	6.300	7.800	9.150	10.95	15.15
17	6.318	7.800	9.212	11.05	15.28
18	6.333	7.733	9.200	10.93	15.27
19	6.347	7.863	9.253	11.02	15.44
20	6.240	7.800	9.240	11.10	15.36
∞	6.251	7.815	9.348	11.34	16.27

$k = 5$

P	10	5	2.5	1	0.1
$n = 3$	7.467	8.533	9.600	10.13	11.47
4	7.600	8.800	9.800	11.20	13.20
5	7.680	8.960	10.24	11.68	14.40
6	7.733	9.067	10.40	11.87	15.20
7	7.771	9.143	10.51	12.11	15.66
8	7.700	9.200	10.60	12.30	16.00
9	7.733	9.244	10.67	12.44	16.36
∞	7.779	9.488	11.14	13.28	18.47

$k = 6$

P	10	5	2.5	1	0.1
$n = 3$	8.714	9.857	10.81	11.76	13.29
4	9.000	10.29	11.43	12.71	15.29
5	9.000	10.49	11.74	13.23	16.43
6	9.048	10.57	12.00	13.62	17.05
∞	9.236	11.07	12.83	15.09	20.52

TABLE 25. UPPER PERCENTAGE POINTS OF THE KRUSKAL–WALLIS DISTRIBUTION

Consider k random samples of sizes $n_1, n_2, ..., n_k$ respectively, $n_1 \geqslant n_2 \geqslant ... \geqslant n_k$, and let $N = n_1 + n_2 + ... + n_k$. Let all the N observations be ranked in increasing order of size, and let R_j ($j = 1, 2, ..., k$) denote the sum of the ranks of the observations belonging to the jth sample. This table gives percentage points of the Kruskal–Wallis statistic

$$H = \frac{12}{N(N+1)} \sum_{j=1}^{k} \frac{R_j^2}{n_j} - 3(N+1)$$

on the assumption that all k samples are from the same continuous population; the function tabulated, $x(P)$, is the smallest x such that, on this assumption, $\Pr\{H \geqslant x\} \leqslant P/100$. A dash indicates that there is no value with the required property. The limiting distribution of H as N tends to infinity and each ratio n_j/N tends to a positive number is the χ^2-distribution with $k-1$ degrees of freedom (see Table 8), and the percentage points are given under $n_j = \infty$ ($j = 1, 2, ..., k$). A test of the hypothesis that all k samples are from the same continuous population is provided by rejecting at the P per cent level if $H \geqslant x(P)$.

k = 3

n_1, n_2, n_3	$P = 10$	5	2.5	1	0.1
2, 2, 2	4.571	—	—	—	—
3, 2, 1	4.286	—	—	—	—
3, 2, 2	4.500	4.714	—	—	—
3, 3, 1	4.571	5.143	—	—	—
3, 3, 2	4.556	5.361	5.556	—	—
3, 3, 3	4.622	5.600	5.956	7.200	—
4, 2, 1	4.500	—	—	—	—
4, 2, 2	4.458	5.333	5.500	—	—
4, 3, 1	4.056	5.208	5.833	—	—
4, 3, 2	4.511	5.444	6.000	6.444	—
4, 3, 3	4.709	5.791	6.155	6.745	—
4, 4, 1	4.167	4.967	6.167	6.667	—
4, 4, 2	4.555	5.455	6.327	7.036	—
4, 4, 3	4.545	5.598	6.394	7.144	8.909
4, 4, 4	4.654	5.692	6.615	7.654	9.269
5, 2, 1	4.200	5.000	—	—	—
5, 2, 2	4.373	5.160	6.000	6.533	—
5, 3, 1	4.018	4.960	6.044	—	—
5, 3, 2	4.651	5.251	6.004	6.909	—
5, 3, 3	4.533	5.648	6.315	7.079	8.727
5, 4, 1	3.987	4.985	5.858	6.955	—
5, 4, 2	4.541	5.273	6.068	7.205	8.591
5, 4, 3	4.549	5.656	6.410	7.445	8.795
5, 4, 4	4.668	5.657	6.673	7.760	9.168
5, 5, 1	4.109	5.127	6.000	7.309	—
5, 5, 2	4.623	5.338	6.346	7.338	8.938
5, 5, 3	4.545	5.705	6.549	7.578	9.284
5, 5, 4	4.523	5.666	6.760	7.823	9.606
5, 5, 5	4.560	5.780	6.740	8.000	9.920
6, 2, 1	4.200	4.822	5.600	—	—
6, 2, 2	4.545	5.345	5.745	6.655	—
6, 3, 1	3.909	4.855	5.945	6.873	—
6, 3, 2	4.682	5.348	6.136	6.970	—
6, 3, 3	4.590	5.615	6.436	7.410	8.692
6, 4, 1	4.038	4.947	5.856	7.106	—
6, 4, 2	4.494	5.340	6.186	7.340	8.827
6, 4, 3	4.604	5.610	6.538	7.500	9.170
6, 4, 4	4.595	5.681	6.667	7.795	9.681
6, 5, 1	4.128	4.990	5.951	7.182	—
6, 5, 2	4.596	5.338	6.196	7.376	9.189
6, 5, 3	4.535	5.602	6.667	7.590	9.669
6, 5, 4	4.522	5.661	6.750	7.936	9.961

k = 3

n_1, n_2, n_3	$P = 10$	5	2.5	1	0.1
6, 5, 5	4.547	5.729	6.788	8.028	10.29
6, 6, 1	4.000	4.945	5.923	7.121	9.692
6, 6, 2	4.438	5.410	6.210	7.467	9.752
6, 6, 3	4.558	5.625	6.725	7.725	10.15
6, 6, 4	4.548	5.724	6.812	8.000	10.34
6, 6, 5	4.542	5.765	6.848	8.124	10.52
6, 6, 6	4.643	5.801	6.889	8.222	10.89
7, 1, 1	4.267	—	—	—	—
7, 2, 1	4.200	4.706	5.727	—	—
7, 2, 2	4.526	5.143	5.818	7.000	—
7, 3, 1	4.173	4.952	5.758	7.030	—
7, 3, 2	4.582	5.357	6.201	6.839	8.654
7, 3, 3	4.603	5.620	6.449	7.228	9.262
7, 4, 1	4.121	4.986	5.791	6.986	—
7, 4, 2	4.549	5.376	6.184	7.321	9.198
7, 4, 3	4.527	5.623	6.578	7.550	9.670
7, 4, 4	4.562	5.650	6.707	7.814	9.841
7, 5, 1	4.035	5.064	5.953	7.061	9.178
7, 5, 2	4.485	5.393	6.221	7.450	9.640
7, 5, 3	4.535	5.607	6.627	7.697	9.874
7, 5, 4	4.542	5.733	6.738	7.931	10.16
7, 5, 5	4.571	5.708	6.835	8.108	10.45
7, 6, 1	4.033	5.067	6.067	7.254	9.747
7, 6, 2	4.500	5.357	6.223	7.490	10.06
7, 6, 3	4.550	5.689	6.694	7.756	10.26
7, 6, 4	4.562	5.706	6.787	8.039	10.46
7, 6, 5	4.560	5.770	6.857	8.157	10.75
7, 6, 6	4.530	5.730	6.897	8.257	11.00
7, 7, 1	3.986	4.986	6.057	7.157	9.871
7, 7, 2	4.491	5.398	6.328	7.491	10.24
7, 7, 3	4.613	5.688	6.708	7.810	10.45
7, 7, 4	4.563	5.766	6.788	8.142	10.69
7, 7, 5	4.546	5.746	6.886	8.257	10.92
7, 7, 6	4.568	5.793	6.927	8.345	11.13
7, 7, 7	4.594	5.818	6.954	8.378	11.32
8, 1, 1	4.418	—	—	—	—
8, 2, 1	4.011	4.909	5.420	—	—
8, 2, 2	4.587	5.356	5.817	6.663	—
8, 3, 1	4.010	4.881	6.064	6.804	—
8, 3, 2	4.451	5.316	6.195	7.022	8.791
8, 3, 3	4.543	5.617	6.588	7.350	9.426
8, 4, 1	4.038	5.044	5.885	6.973	8.901
8, 4, 2	4.500	5.393	6.193	7.350	9.293

TABLE 25. UPPER PERCENTAGE POINTS OF THE KRUSKAL–WALLIS DISTRIBUTION

$k = 3$

n_1, n_2, n_3	$P = 10$	5	2.5	1	0.1
8, 4, 3	4.529	5.623	6.562	7.585	9.742
8, 4, 4	4.561	5.779	6.750	7.853	10.01
8, 5, 1	3.967	4.869	5.864	7.110	9.579
8, 5, 2	4.466	5.415	6.260	7.440	9.781
8, 5, 3	4.514	5.614	6.614	7.706	10.04
8, 5, 4	4.549	5.718	6.782	7.992	10.29
8, 5, 5	4.555	5.769	6.843	8.116	10.64
8, 6, 1	4.015	5.015	5.933	7.256	9.840
8, 6, 2	4.463	5.404	6.294	7.522	10.11
8, 6, 3	4.575	5.678	6.658	7.796	10.37
8, 6, 4	4.563	5.743	6.795	8.045	10.63
8, 6, 5	4.550	5.750	6.867	8.226	10.89
8, 6, 6	4.599	5.770	6.932	8.313	11.10
8, 7, 1	4.045	5.041	6.047	7.308	10.03
8, 7, 2	4.451	5.403	6.339	7.571	10.36
8, 7, 3	4.556	5.698	6.671	7.827	10.54
8, 7, 4	4.548	5.759	6.837	8.118	10.84
8, 7, 5	4.551	5.782	6.884	8.242	11.03
8, 7, 6	4.553	5.781	6.917	8.333	11.28
8, 7, 7	4.585	5.802	6.980	8.363	11.42
8, 8, 1	4.044	5.039	6.005	7.314	10.16
8, 8, 2	4.509	5.408	6.351	7.654	10.46
8, 8, 3	4.555	5.734	6.682	7.889	10.69
8, 8, 4	4.579	5.743	6.886	8.168	10.97
8, 8, 5	4.573	5.761	6.920	8.297	11.18
8, 8, 6	4.572	5.779	6.953	8.367	11.37
8, 8, 7	4.571	5.791	6.980	8.419	11.55
8, 8, 8	4.595	5.805	6.995	8.465	11.70
9, 9, 9	4.582	5.845	7.041	8.564	11.95
∞, ∞, ∞	4.605	5.991	7.378	9.210	13.82

$k = 4$

n_1, n_2, n_3, n_4	$P = 10$	5	2.5	1	0.1
2, 2, 2, 1	5.357	5.679	—	—	—
2, 2, 2, 2	5.667	6.167	6.667	6.667	—
3, 2, 1, 1	5.143	—	—	—	—
3, 2, 2, 1	5.556	5.833	6.250	—	—
3, 2, 2, 2	5.644	6.333	6.978	7.133	—
3, 3, 1, 1	5.333	6.333	6.333	—	—
3, 3, 2, 1	5.689	6.244	6.689	7.200	—
3, 3, 2, 2	5.745	6.527	7.055	7.636	8.455
3, 3, 3, 1	5.655	6.600	7.036	7.400	—
3, 3, 3, 2	5.879	6.727	7.515	8.015	9.030
3, 3, 3, 3	6.026	7.000	7.667	8.538	9.513
4, 2, 1, 1	5.250	5.833	—	—	—
4, 2, 2, 1	5.533	6.133	6.533	7.000	—
4, 2, 2, 2	5.755	6.545	7.064	7.391	—
4, 3, 1, 1	5.067	6.178	6.711	7.067	—
4, 3, 2, 1	5.591	6.309	6.955	7.455	—
4, 3, 2, 2	5.750	6.621	7.326	7.871	8.909
4, 3, 3, 1	5.689	6.545	7.326	7.758	9.182
4, 3, 3, 2	5.872	6.795	7.564	8.333	9.455
4, 3, 3, 3	6.016	6.984	7.775	8.659	10.02
4, 4, 1, 1	5.182	5.945	6.955	7.909	—
4, 4, 2, 1	5.568	6.386	7.159	7.909	8.909
4, 4, 2, 2	5.808	6.731	7.538	8.346	9.462
4, 4, 3, 1	5.692	6.635	7.500	8.231	9.327
4, 4, 3, 2	5.901	6.874	7.747	8.621	9.945
4, 4, 3, 3	6.019	7.038	7.929	8.876	10.47
4, 4, 4, 1	5.654	6.725	7.648	8.588	9.758
4, 4, 4, 2	5.914	6.957	7.914	8.871	10.43
4, 4, 4, 3	6.042	7.142	8.079	9.075	10.93
4, 4, 4, 4	6.088	7.235	8.228	9.287	11.36
$\infty, \infty, \infty, \infty$	6.251	7.815	9.348	11.34	16.27

$k = 5$

n_1, n_2, n_3, n_4, n_5	$P = 10$	5	2.5	1	0.1
2, 2, 1, 1, 1	5.786	—	—	—	—
2, 2, 2, 1, 1	6.250	6.750	6.750	—	—
2, 2, 2, 2, 1	6.600	7.133	7.333	7.533	—
2, 2, 2, 2, 2	6.982	7.418	7.964	8.291	—
3, 2, 1, 1, 1	6.139	6.583	—	—	—
3, 2, 2, 1, 1	6.511	6.800	7.200	7.600	—
3, 2, 2, 2, 1	6.709	7.309	7.745	8.127	—
3, 2, 2, 2, 2	6.955	7.682	8.182	8.682	9.364
3, 3, 1, 1, 1	6.311	7.111	7.467	—	—
3, 3, 2, 1, 1	6.600	7.200	7.618	8.073	—
3, 3, 2, 2, 1	6.788	7.591	8.121	8.576	9.303
3, 3, 2, 2, 2	7.026	7.910	8.538	9.115	10.03
3, 3, 3, 1, 1	6.788	7.576	8.061	8.424	9.455
3, 3, 3, 2, 1	6.910	7.769	8.449	9.051	9.974
3, 3, 3, 2, 2	7.121	8.044	8.813	9.505	10.64
3, 3, 3, 3, 1	7.077	8.000	8.703	9.451	10.59
3, 3, 3, 3, 2	7.210	8.200	9.038	9.876	11.17
3, 3, 3, 3, 3	7.333	8.333	9.233	10.20	11.67
$\infty, \infty, \infty, \infty, \infty$	7.779	9.488	11.14	13.28	18.47

$k = 6$

$n_1, n_2, n_3, n_4, n_5, n_6$	$P = 10$	5	2.5	1	0.1
2, 2, 1, 1, 1, 1	6.833	—	—	—	—
2, 2, 2, 1, 1, 1	7.267	7.600	7.800	—	—
2, 2, 2, 2, 1, 1	7.527	8.018	8.345	8.618	—
2, 2, 2, 2, 2, 1	7.909	8.455	8.864	9.227	9.773
2, 2, 2, 2, 2, 2	8.154	8.846	9.385	9.846	10.54
3, 2, 1, 1, 1, 1	7.133	7.467	7.667	—	—
3, 2, 2, 1, 1, 1	7.418	7.945	8.236	8.509	—
3, 2, 2, 2, 1, 1	7.727	8.348	8.727	9.136	9.682
3, 2, 2, 2, 2, 1	7.987	8.731	9.218	9.692	10.38
3, 2, 2, 2, 2, 2	8.198	9.033	9.648	10.22	11.11
3, 3, 1, 1, 1, 1	7.400	7.909	8.564	8.564	—
3, 3, 2, 1, 1, 1	7.697	8.303	8.667	9.045	—
3, 3, 2, 2, 1, 1	7.872	8.615	9.128	9.628	10.31
3, 3, 2, 2, 2, 1	8.077	8.923	9.549	10.15	11.01
3, 3, 2, 2, 2, 2	8.305	9.190	9.914	10.61	11.68
$\infty, \infty, \infty, \infty, \infty, \infty$	9.236	11.07	12.83	15.09	20.52

TABLE 26. HYPERGEOMETRIC PROBABILITIES

Suppose that of N objects, R are of type A and $N-R$ of type B, with $R \leqslant N-R$. Suppose that n of the objects, $n \leqslant N-n$, are selected at random without replacement and X are found to be of type A. Then X follows a hypergeometric distribution with the probability that $X = r$ given by

$$p(r|N, R, n) = \binom{R}{r}\binom{N-R}{n-r} \Big/ \binom{N}{n}.$$

This table gives these probabilities for $N \leqslant 17$ and $n \leqslant R$ (if not, use the result that $p(r|N, R, n) = p(r|N, n, R)$). For $N > 17$ these probabilities may be calculated by using binomial coefficients (Table 3) or logarithms of factorials (Table 6). When N is large and $R/N < 0\cdot1$, X is approximately binomially distributed with index R and parameter $p = n/N$ (see Table 1); similarly, if N is large and $n/N < 0\cdot1$, X is approximately binomially distributed with index n and parameter $p = R/N$. If N is large and neither R/N nor n/N is less than $0\cdot1$,

$$(X + \tfrac{1}{2} - nR/N)/[R(N-R)\, n(N-n)/N^2(N-1)]^{\frac{1}{2}}$$

is approximately normally distributed with zero mean and unit variance; a continuity correction of $\frac{1}{2}$, as with the binomial distribution, has been used.

A representation of the data in the form of a 2×2 contingency table is useful:

r	$R-r$	R
$n-r$	$N-R-n+r$	$N-R$
n	$N-n$	N

Here the rows correspond to types A and B and the columns to 'selected' and 'not selected' respectively, and the marginal totals are given.

Fisher's exact test of no association between rows and columns, or of homogeneity of types A and B, is provided by rejecting the null hypothesis at the P per cent level if the sum of the probabilities for *all* tables with at least as extreme values of X as that observed is less than or equal to $P/100$. 'More extreme' means having smaller probability than the observed value r of X, given the same marginal totals. This test may be either one- or two-sided, as shown below.

Example.

1	5	6
4	4	8
5	9	14

From the tables $p(1|14, 6, 5) = \cdot2098$. A more extreme one-sided value is $r = 0$, giving a total probability of $\cdot2378$, not significant evidence of association or of inhomogeneity. If a two-sided test is required, $r = 4$ and $r = 5$ have probabilities $\cdot0599$ and $\cdot0030$ respectively, less than $\cdot2098$; the total is now $\cdot3007$.

When $N > 17$ and nR/N is not too small, a (two-sided) test of the hypothesis of no association, or of homogeneity, is provided by rejecting at the P per cent level approximately if $\chi^2 = N[rN-nR]^2/[R(N-R)\, n(N-n)]$ exceeds $\chi_1^2(P)$ (see Table 8). (Cf. H. Cramér, *Mathematical Methods of Statistics* (1946), Princeton University Press, Princeton, N.J., Sections 30.5 and 30.6.)

$N\ R\ n$ = 2 1 1

r	p
$r = 0$	0·5000
1	·5000

3 1 1

r	p
0	0·6667
1	·3333

4 1 1

r	p
0	0·7500
1	·2500

4 2 1

r	p
0	0·5000
1	·5000

4 2 2

r	p
0	0·1667
1	·6667
2	·1667

5 1 1

r	p
0	0·8000
1	·2000

5 2 1

r	p
0	0·6000
1	·4000

---

5 2 2

r	p
$r = 0$	0·3000
1	·6000
2	·1000

6 1 1

r	p
0	0·8333
1	·1667

6 2 1

r	p
0	0·6667
1	·3333

6 2 2

r	p
0	0·4000
1	·5333
2	·0667

6 3 1

r	p
0	0·5000
1	·5000

6 3 2

r	p
0	0·2000
1	·6000
2	·2000

6 3 3

r	p
0	0·0500
1	·4500
2	·4500
3	·0500

---

7 1 1

r	p
$r = 0$	0·8571
1	·1429

7 2 1

r	p
0	0·7143
1	·2857

7 2 2

r	p
0	0·4762
1	·4762
2	·0476

7 3 1

r	p
0	0·5714
1	·4286

7 3 2

r	p
0	0·2857
1	·5714
2	·1429

7 3 3

r	p
0	0·1143
1	·5143
2	·3429
3	·0286

8 1 1

r	p
0	0·8750
1	·1250

---

8 2 1

r	p
$r = 0$	0·7500
1	·2500

8 2 2

r	p
0	0·5357
1	·4286
2	·0357

8 3 1

r	p
0	0·6250
1	·3750

8 3 2

r	p
0	0·3571
1	·5357
2	·1071

8 3 3

r	p
0	0·1786
1	·5357
2	·2679
3	·0179

8 4 1

r	p
0	0·5000
1	·5000

8 4 2

r	p
0	0·2143
1	·5714
2	·2143

---

8 4 3

r	p
$r = 0$	0·0714
1	·4286
2	·4286
3	·0714

8 4 4

r	p
0	0·0143
1	·2286
2	·5143
3	·2286
4	·0143

9 1 1

r	p
0	0·8889
1	·1111

9 2 1

r	p
0	0·7778
1	·2222

9 2 2

r	p
0	0·5833
1	·3889
2	·0278

9 3 1

r	p
0	0·6667
1	·3333

---

9 3 2

r	p
$r = 0$	0·4167
1	·5000
2	·0833

9 3 3

r	p
0	0·2381
1	·5357
2	·2143
3	·0119

9 4 1

r	p
0	0·5556
1	·4444

9 4 2

r	p
0	0·2778
1	·5556
2	·1667

9 4 3

r	p
0	0·1190
1	·4762
2	·3571
3	·0476

9 4 4

r	p
0	0·0397
1	·3175
2	·4762
3	·1587
4	·0079

---

10 1 1

r	p
$r = 0$	0·9000
1	·1000

10 2 1

r	p
0	0·8000
1	·2000

10 2 2

r	p
0	0·6222
1	·3556
2	·0222

10 3 1

r	p
0	0·7000
1	·3000

10 3 2

r	p
0	0·4667
1	·4667
2	·0667

10 3 3

r	p
0	0·2917
1	·5250
2	·1750
3	·0083

10 4 1

r	p
0	0·6000
1	·4000

TABLE 26. HYPERGEOMETRIC PROBABILITIES

N R n			N R n			N R n			N R n			N R n			N R n			N R n		
10	4	2	11	3	2	12	1	1	12	5	4	13	2	2	13	5	5	14	3	2
$r=0$	0·3333		$r=0$	0·5091		$r=0$	0·9167		$r=0$	0·0707		$r=0$	0·7051		$r=0$	0·0435		$r=0$	0·6044	
1	·5333		1	·4364		1	·0833		1	·3535		1	·2821		1	·2720		1	·3626	
2	·1333		2	·0545					2	·4242		2	·0128		2	·4351		2	·0330	
						12	2	1	3	·1414					3	·2176				
10	4	3	11	3	3	0	0·8333		4	·0101		13	3	1	4	·0311		14	3	3
0	0·1667		0	0·3394		1	·1667					0	0·7692		5	·0008		0	0·4533	
1	·5000		1	·5091					12	5	5	1	·2308					1	·4533	
2	·3000		2	·1455		12	2	2	0	0·0265					13	6	1	2	·0907	
3	·0333		3	·0061		0	0·6818		1	·2210		13	3	2	0	0·5385		3	·0027	
						1	·3030		2	·4419		0	0·5769		1	·4615				
10	4	4	11	4	1	2	·0152		3	·2652		1	·3846					14	4	1
0	0·0714		0	0·6364					4	·0442		2	·0385		13	6	2	0	0·7143	
1	·3810		1	·3636		12	3	1	5	·0013					0	0·2692		1	·2857	
2	·4286					0	0·7500					13	3	3	1	·5385				
3	·1143		11	4	2	1	·2500		12	6	1	0	0·4196		2	·1923		14	4	2
4	·0048		0	0·3818					0	0·5000		1	·4720					0	0·4945	
			1	·5091		12	3	2	1	·5000		2	·1049		13	6	3	1	·4396	
10	5	1	2	·1091		0	0·5455					3	·0035		0	0·1224		2	·0659	
0	0·5000					1	·4091		12	6	2				1	·4406				
1	·5000		11	4	3	2	·0455		0	0·2273		13	4	1	2	·3671		14	4	3
			0	0·2121					1	·5455		0	0·6923		3	·0699		0	0·3297	
10	5	2	1	·5091		12	3	3	2	·2273		1	·3077					1	·4945	
0	0·2222		2	·2545		0	0·3818								13	6	4	2	·1648	
1	·5556		3	·0242		1	·4909		12	6	3	13	4	2	0	0·0490		3	·0110	
2	·2222					2	·1227		0	0·0909		0	0·4615		1	·2937				
			11	4	4	3	·0045		1	·4091		1	·4615		2	·4406		14	4	4
10	5	3	0	0·1061					2	·4091		2	·0769		3	·1958		0	0·2098	
0	0·0833		1	·4242		12	4	1	3	·0909					4	·0210		1	·4795	
1	·4167		2	·3818		0	0·6667					13	4	3				2	·2697	
2	·4167		3	·0848		1	·3333		12	6	4	0	0·2937		13	6	5	3	·0400	
3	·0833		4	·0030					0	0·0303		1	·5035		0	0·0163		4	·0010	
						12	4	2	1	·2424		2	·1888		1	·1632				
10	5	4	11	5	1	0	0·4242		2	·4545		3	·0140		2	·4079		14	5	1
0	0·0238		0	0·5455		1	·4848		3	·2424					3	·3263		0	0·6429	
1	·2381		1	·4545		2	·0909		4	·0303		13	4	4	4	·0816		1	·3571	
2	·4762											0	0·1762		5	·0047				
3	·2381		11	5	2	12	4	3	12	6	5	1	·4699					14	5	2
4	·0238		0	0·2727		0	0·2545		0	0·0076		2	·3021		13	6	6	0	0·3956	
			1	·5455		1	·5091		1	·1136		3	·0503		0	0·0041		1	·4945	
10	5	5	2	·1818		2	·2182		2	·3788		4	·0014		1	·0734		2	·1099	
0	0·0040					3	·0182		3	·3788					2	·3059				
1	·0992		11	5	3				4	·1136		13	5	1	3	·4079		14	5	3
2	·3968		0	0·1212		12	4	4	5	·0076		0	0·6154		4	·1836		0	0·2308	
3	·3968		1	·4545		0	0·1414					1	·3846		5	·0245		1	·4945	
4	·0992		2	·3636		1	·4525		12	6	6				6	·0006		2	·2473	
5	·0040		3	·0606		2	·3394		0	0·0011		13	5	2				3	·0275	
						3	·0646		1	·0390		0	0·3590		14	1	1			
			11	5	4	4	·0020		2	·2435		1	·5128		0	0·9286		14	5	4
11	1	1	0	0·0455					3	·4329		2	·1282		1	·0714		0	0·1259	
0	0·9091		1	·3030		12	5	1	4	·2435								1	·4196	
1	·0909		2	·4545		0	0·5833		5	·0390		13	5	3	14	2	1	2	·3596	
			3	·1818		1	·4167		6	·0011		0	0·1958		0	0·8571		3	·0899	
11	2	1	4	·0152								1	·4895		1	·1429		4	·0050	
0	0·8182					12	5	2	13	1	1	2	·2797							
1	·1818		11	5	5	0	0·3182		0	0·9231		3	·0350		14	2	2	14	5	5
			0	0·0130		1	·5303		1	·0769					0	0·7253		0	0·0629	
11	2	2	1	·1623		2	·1515					13	5	4	1	·2637		1	·3147	
0	0·6545		2	·4329					13	2	1	0	0·0979		2	·0110		2	·4196	
1	·3273		3	·3247		12	5	3	0	0·8462		1	·3916					3	·1798	
2	·0182		4	·0649		0	0·1591		1	·1538		2	·3916		14	3	1	4	·0225	
			5	·0022		1	·4773					3	·1119		0	0·7857		5	·0005	
11	3	1				2	·3182					4	·0070		1	·2143				
0	0·7273					3	·0455													
1	·2727																			

TABLE 26. HYPERGEOMETRIC PROBABILITIES

N	R	n
14	6	1

r		
0	0.5714	
1	.4286	

N	R	n
14	6	2

0	0.3077	
1	.5275	
2	.1648	

N	R	n
14	6	3

0	0.1538	
1	.4615	
2	.3297	
3	.0549	

N	R	n
14	6	4

0	0.0699	
1	.3357	
2	.4196	
3	.1598	
4	.0150	

N	R	n
14	6	5

0	0.0280	
1	.2098	
2	.4196	
3	.2797	
4	.0599	
5	.0030	

N	R	n
14	6	6

0	0.0093	
1	.1119	
2	.3497	
3	.3730	
4	.1399	
5	.0160	
6	.0003	

N	R	n
14	7	1

0	0.5000	
1	.5000	

N	R	n
14	7	2

0	0.2308	
1	.5385	
2	.2308	

N	R	n
14	7	3

0	0.0962	
1	.4038	
2	.4038	
3	.0962	

N	R	n
14	7	4

0	0.0350	
1	.2448	
2	.4406	
3	.2448	
4	.0350	

N	R	n
14	7	5

r		
0	0.0105	
1	.1224	
2	.3671	
3	.3671	
4	.1224	
5	.0105	

N	R	n
14	7	6

0	0.0023	
1	.0490	
2	.2448	
3	.4079	
4	.2448	
5	.0490	
6	.0023	

N	R	n
14	7	7

0	0.0003	
1	.0143	
2	.1285	
3	.3569	
4	.3569	
5	.1285	
6	.0143	
7	.0003	

N	R	n
15	1	1

0	0.9333	
1	.0667	

N	R	n
15	2	1

0	0.8667	
1	.1333	

N	R	n
15	2	2

0	0.7429	
1	.2476	
2	.0095	

N	R	n
15	3	1

0	0.8000	
1	.2000	

N	R	n
15	3	2

0	0.6286	
1	.3429	
2	.0286	

N	R	n
15	3	3

0	0.4835	
1	.4352	
2	.0791	
3	.0022	

N	R	n
15	4	1

0	0.7333	
1	.2667	

N	R	n
15	4	2

0	0.5238	
1	.4190	
2	.0571	

N	R	n
15	4	3

r		
0	0.3626	
1	.4835	
2	.1451	
3	.0088	

N	R	n
15	4	4

0	0.2418	
1	.4835	
2	.2418	
3	.0322	
4	.0007	

N	R	n
15	5	1

0	0.6667	
1	.3333	

N	R	n
15	5	2

0	0.4286	
1	.4762	
2	.0952	

N	R	n
15	5	3

0	0.2637	
1	.4945	
2	.2198	
3	.0220	

N	R	n
15	5	4

0	0.1538	
1	.4396	
2	.3297	
3	.0733	
4	.0037	

N	R	n
15	5	5

0	0.0839	
1	.3497	
2	.3996	
3	.1499	
4	.0167	
5	.0003	

N	R	n
15	6	1

0	0.6000	
1	.4000	

N	R	n
15	6	2

0	0.3429	
1	.5143	
2	.1429	

N	R	n
15	6	3

0	0.1846	
1	.4747	
2	.2967	
3	.0440	

N	R	n
15	6	4

0	0.0923	
1	.3692	
2	.3956	
3	.1319	
4	.0110	

N	R	n
15	6	5

r		
0	0.0420	
1	.2517	
2	.4196	
3	.2398	
4	.0450	
5	.0020	

N	R	n
15	6	6

0	0.0168	
1	.1510	
2	.3776	
3	.3357	
4	.1079	
5	.0108	
6	.0002	

N	R	n
15	7	1

0	0.5333	
1	.4667	

N	R	n
15	7	2

0	0.2667	
1	.5333	
2	.2000	

N	R	n
15	7	3

0	0.1231	
1	.4308	
2	.3692	
3	.0769	

N	R	n
15	7	4

0	0.0513	
1	.2872	
2	.4308	
3	.2051	
4	.0256	

N	R	n
15	7	5

0	0.0186	
1	.1632	
2	.3916	
3	.3263	
4	.0932	
5	.0070	

N	R	n
15	7	6

0	0.0056	
1	.0783	
2	.2937	
3	.3916	
4	.1958	
5	.0336	
6	.0014	

N	R	n
15	7	7

0	0.0012	
1	.0305	
2	.1828	
3	.3807	
4	.3046	
5	.0914	
6	.0087	
7	.0002	

N	R	n
16	1	1

r		
0	0.9375	
1	.0625	

N	R	n
16	2	1

0	0.8750	
1	.1250	

N	R	n
16	2	2

0	0.7583	
1	.2333	
2	.0083	

N	R	n
16	3	1

0	0.8125	
1	.1875	

N	R	n
16	3	2

0	0.6500	
1	.3250	
2	.0250	

N	R	n
16	3	3

0	0.5107	
1	.4179	
2	.0696	
3	.0018	

N	R	n
16	4	1

0	0.7500	
1	.2500	

N	R	n
16	4	2

0	0.5500	
1	.4000	
2	.0500	

N	R	n
16	4	3

0	0.3929	
1	.4714	
2	.1286	
3	.0071	

N	R	n
16	4	4

0	0.2720	
1	.4835	
2	.2176	
3	.0264	
4	.0005	

N	R	n
16	5	1

0	0.6875	
1	.3125	

N	R	n
16	5	2

0	0.4583	
1	.4583	
2	.0833	

N	R	n
16	5	3

0	0.2946	
1	.4911	
2	.1964	
3	.0179	

N	R	n
16	5	4

r		
0	0.1813	
1	.4533	
2	.3022	
3	.0604	
4	.0027	

N	R	n
16	5	5

0	0.1058	
1	.3777	
2	.3777	
3	.1259	
4	.0126	
5	.0002	

N	R	n
16	6	1

0	0.6250	
1	.3750	

N	R	n
16	6	2

0	0.3750	
1	.5000	
2	.1250	

N	R	n
16	6	3

0	0.2143	
1	.4821	
2	.2679	
3	.0357	

N	R	n
16	6	4

0	0.1154	
1	.3956	
2	.3709	
3	.1099	
4	.0082	

N	R	n
16	6	5

0	0.0577	
1	.2885	
2	.4121	
3	.2060	
4	.0343	
5	.0014	

N	R	n
16	6	6

0	0.0262	
1	.1888	
2	.3934	
3	.2997	
4	.0843	
5	.0075	
6	.0001	

N	R	n
16	7	1

0	0.5625	
1	.4375	

N	R	n
16	7	2

0	0.3000	
1	.5250	
2	.1750	

N	R	n
16	7	3

r		
0	0.1500	
1	.4500	
2	.3375	
3	.0625	

N	R	n
16	7	4

0	0.0692	
1	.3231	
2	.4154	
3	.1731	
4	.0192	

N	R	n
16	7	5

0	0.0288	
1	.2019	
2	.4038	
3	.2885	
4	.0721	
5	.0048	

N	R	n
16	7	6

0	0.0105	
1	.1101	
2	.3304	
3	.3671	
4	.1573	
5	.0236	
6	.0009	

N	R	n
16	7	7

0	0.0031	
1	.0514	
2	.2313	
3	.3855	
4	.2570	
5	.0661	
6	.0055	
7	.0001	

N	R	n
16	8	1

0	0.5000	
1	.5000	

N	R	n
16	8	2

0	0.2333	
1	.5333	
2	.2333	

N	R	n
16	8	3

0	0.1000	
1	.4000	
2	.4000	
3	.1000	

N	R	n
16	8	4

0	0.0385	
1	.2462	
2	.4308	
3	.2462	
4	.0385	

TABLE 26. HYPERGEOMETRIC PROBABILITIES

N R n = 16 8 5

r	
0	0·0128
1	·1282
2	·3590
3	·3590
4	·1282
5	·0128

N R n = 16 8 6

r	
0	0·0035
1	·0559
2	·2448
3	·3916
4	·2448
5	·0559
6	·0035

N R n = 16 8 7

r	
0	0·0007
1	·0196
2	·1371
3	·3427
4	·3427
5	·1371
6	·0196
7	·0007

N R n = 16 8 8

r	
0	0·0001
1	·0050
2	·0609
3	·2437
4	·3807
5	·2437
6	·0609
7	·0050
8	·0001

N R n = 17 1 1

r	
0	0·9412
1	·0588

N R n = 17 2 1

r	
0	0·8824
1	·1176

N R n = 17 2 2

r	
0	0·7721
1	·2206
2	·0074

N R n = 17 3 1

r	
0	0·8235
1	·1765

N R n = 17 3 2

r	
0	0·6691
1	·3088
2	·0221

N R n = 17 3 3

r	
0	0·5353
1	·4015
2	·0618
3	·0015

N R n = 17 4 1

r	
0	0·7647
1	·2353

N R n = 17 4 2

r	
0	0·5735
1	·3824
2	·0441

N R n = 17 4 3

r	
0	0·4206
1	·4588
2	·1147
3	·0059

N R n = 17 4 4

r	
0	0·3004
1	·4807
2	·1966
3	·0218
4	·0004

N R n = 17 5 1

r	
0	0·7059
1	·2941

N R n = 17 5 2

r	
0	0·4853
1	·4412
2	·0735

N R n = 17 5 3

r	
0	0·3235
1	·4853
2	·1765
3	·0147

N R n = 17 5 4

r	
0	0·2080
1	·4622
2	·2773
3	·0504
4	·0021

N R n = 17 5 5

r	
0	0·1280
1	·4000
2	·3555
3	·1067
4	·0097
5	·0002

N R n = 17 6 1

r	
0	0·6471
1	·3529

N R n = 17 6 2

r	
0	0·4044
1	·4853
2	·1103

N R n = 17 6 3

r	
0	0·2426
1	·4853
2	·2426
3	·0294

N R n = 17 6 4

r	
0	0·1387
1	·4160
2	·3466
3	·0924
4	·0063

N R n = 17 6 5

r	
0	0·0747
1	·3200
2	·4000
3	·1778
4	·0267
5	·0010

N R n = 17 6 6

r	
0	0·0373
1	·2240
2	·4000
3	·2666
4	·0667
5	·0053
6	·0001

N R n = 17 7 1

r	
0	0·5882
1	·4118

N R n = 17 7 2

r	
0	0·3309
1	·5147
2	·1544

N R n = 17 7 3

r	
0	0·1765
1	·4632
2	·3088
3	·0515

N R n = 17 7 4

r	
0	0·0882
1	·3529
2	·3971
3	·1471
4	·0147

N R n = 17 7 5

r	
0	0·0407
1	·2376
2	·4072
3	·2545
4	·0566
5	·0034

N R n = 17 7 6

r	
0	0·0170
1	·1425
2	·3563
3	·3394
4	·1273
5	·0170
6	·0006

N R n = 17 7 7

r	
0	0·0062
1	·0756
2	·2721
3	·3779
4	·2160
5	·0486
6	·0036
7	·0001

N R n = 17 8 1

r	
0	0·5294
1	·4706

N R n = 17 8 2

r	
0	0·2647
1	·5294
2	·2059

N R n = 17 8 3

r	
0	0·1235
1	·4235
2	·3706
3	·0824

N R n = 17 8 4

r	
0	0·0529
1	·2824
2	·4235
3	·2118
4	·0294

N R n = 17 8 5

r	
0	0·0204
1	·1629
2	·3801
3	·3258
4	·1018
5	·0090

N R n = 17 8 6

r	
0	0·0068
1	·0814
2	·2851
3	·3801
4	·2036
5	·0407
6	·0023

N R n = 17 8 7

r	
0	0·0019
1	·0346
2	·1814
3	·3628
4	·3023
5	·1037
6	·0130
7	·0004

N R n = 17 8 8

r	
0	0·0004
1	·0118
2	·0968
3	·2903
4	·3628
5	·1935
6	·0415
7	·0030
8	·0000

TABLE 27. RANDOM SAMPLING NUMBERS

Each digit is an independent sample from a population in which the digits 0 to 9 are equally likely, that is, each has a probability of $\frac{1}{10}$.

84	42	56	53	87	75	18	91	76	66	64	83	97	11	69	41	80	92	38	75
28	87	77	03	57	09	85	86	46	86	40	15	31	81	78	91	30	22	88	58
64	12	39	65	37	93	76	46	11	09	56	28	94	54	10	14	30	73	80	30
49	41	73	76	49	64	06	70	99	37	72	60	39	16	02	26	91	90	16	54
06	46	69	31	24	33	52	67	85	07	01	33	16	33	43	98	17	62	52	52
75	56	96	97	65	20	68	68	60	97	90	46	63	37	10	34	41	64	85	01
09	35	89	97	97	10	00	76	39	82	49	94	15	89	60	65	57	03	91	68
73	81	11	08	52	73	64	85	22	72	85	16	15	97	76	28	41	95	00	33
49	69	80	41	46	62	26	32	58	16	88	76	54	32	06	37	46	45	28	95
64	60	49	70	33	73	71	57	83	26	19	25	86	21	64	60	11	01	86	70
93	05	36	44	59	19	99	51	54	21	37	48	18	60	22	92	68	34	39	02
39	88	11	26	68	92	81	14	12	16	37	64	61	48	21	69	77	76	33	00
89	34	19	12	83	76	35	11	96	53	04	76	63	10	93	68	52	42	73	20
77	29	03	26	45	36	15	17	27	28	79	58	38	98	73	52	63	72	48	41
86	75	51	29	70	78	24	78	94	78	64	17	32	23	95	52	87	79	14	30
95	98	77	51	14	65	76	49	42	36	11	33	23	89	32	01	60	48	91	44
22	09	01	14	04	96	97	56	92	52	83	44	45	08	72	78	10	36	26	70
30	49	36	23	36	81	11	76	91	08	67	60	01	15	64	77	21	33	72	29
77	59	88	92	17	75	04	47	18	02	94	84	71	44	87	63	06	04	49	33
03	50	80	26	74	74	18	85	92	20	64	39	98	68	29	26	90	14	77	36
46	32	79	69	41	06	26	04	47	24	67	10	66	69	21	55	66	63	48	47
65	73	98	08	05	96	92	27	22	86	54	87	95	87	40	27	09	97	47	21
68	82	77	73	08	37	28	47	73	49	10	65	53	48	87	74	02	99	52	86
93	98	12	19	82	69	61	08	00	42	88	83	70	85	08	48	74	94	88	61
61	27	39	16	42	17	89	81	27	44	12	33	43	24	92	41	55	13	45	01
54	74	04	79	72	61	21	87	23	83	96	56	97	63	67	02	67	30	36	89
28	00	40	86	92	97	06	22	37	37	83	00	97	17	08	06	43	95	76	84
61	78	71	16	41	01	69	63	35	96	60	65	09	44	93	42	72	11	22	85
68	60	92	99	60	97	53	55	34	61	43	40	77	96	19	87	63	49	22	47
21	76	13	39	25	89	91	38	25	19	44	33	11	36	72	21	40	90	76	95
73	59	53	04	35	13	12	31	88	70	05	40	43	42	47	17	03	86	14	10
85	68	66	48	05	24	28	97	84	84	91	65	62	83	89	68	07	51	01	02
60	30	10	46	44	34	19	56	00	83	20	53	53	05	29	03	47	55	23	26
44	63	80	62	80	80	99	43	33	87	70	52	51	62	02	12	02	90	44	44
89	38	13	68	31	31	97	15	35	67	23	74	76	96	62	82	62	19	65	58
55	20	77	12	79	81	42	15	30	67	88	83	69	08	99	82	20	39	92	40
67	40	42	16	46	06	60	74	61	22	95	47	24	62	81	06	19	67	15	06
57	19	76	98	65	64	55	28	34	03	58	62	35	22	67	40	04	88	17	59
21	72	97	04	82	62	09	54	35	17	22	73	35	72	53	65	95	48	55	12
46	89	95	61	31	77	14	14	24	14	91	58	76	56	19	33	98	67	09	04
99	73	85	64	96	58	61	65	60	83	62	10	87	00	82	63	39	90	83	17
85	52	98	27	40	33	09	59	80	17	22	06	84	03	41	48	76	07	26	69
50	12	17	86	50	57	91	28	42	29	83	87	00	87	93	52	53	47	08	65
92	84	02	93	44	36	93	19	08	54	76	62	31	65	94	68	38	04	62	31
69	74	30	25	68	65	19	77	57	05	71	56	91	30	16	66	70	48	78	65
51	69	76	00	20	92	58	21	24	33	74	08	66	90	61	89	56	83	39	58
27	25	81	29	75	02	85	09	58	89	77	83	03	40	21	14	45	90	54	01
44	03	62	96	68	65	24	57	44	43	07	72	59	16	04	94	23	36	55	85
40	59	49	20	48	63	35	74	33	12	96	25	59	35	07	45	80	97	19	90
92	91	07	14	82	22	50	70	75	15	69	71	31	20	60	06	99	56	57	74

TABLE 28. RANDOM NORMAL DEVIATES

Each number in this table is an independent sample from the normal distribution with zero mean and unit variance.

0·7691	1·0861	−0·9189	−0·1051	−0·7442	−0·2884	−1·4119	0·9222	1·6674	−0·1543
−0·5256	1·5109	−0·5447	−0·2588	1·2474	−1·9211	−1·2664	0·0989	−0·8427	−0·1108
0·9614	0·3639	0·6299	−0·7164	1·1397	0·4393	−0·4255	−0·8744	0·0398	0·5970
0·3003	1·7218	−0·1507	0·1110	1·9225	−0·2911	−0·2201	0·4547	0·5787	0·0842
1·1853	−1·7850	−1·1798	−0·3172	0·8966	1·5256	0·6820	0·0213	−0·7459	−1·5805
0·2411	−0·2628	1·3921	−0·6995	0·8897	1·8266	−0·4070	−0·5601	−0·2542	3·4347
0·2614	0·3413	1·1049	−0·0617	0·4814	0·8452	−0·2685	−0·0945	−0·4762	0·7703
1·0204	0·8185	−0·3797	1·2541	−0·4014	−1·4908	0·7258	−1·1761	0·9149	−0·6135
1·0167	−0·1489	−0·9013	1·4531	−0·9607	0·2071	−1·4880	−0·5156	−0·8178	1·3934
−1·1211	0·4711	1·6169	−0·8250	1·2580	−0·4597	−0·4037	−0·5671	−0·5567	−0·3337
0·2836	0·8620	−0·8698	−0·1144	1·1002	0·1176	−1·1855	−0·5094	0·2013	0·4899
−1·7888	2·4288	0·4890	−0·0339	−0·6501	−0·4560	−0·0251	0·2414	1·3071	0·9586
−0·4654	0·0450	−0·1291	−0·1236	1·7248	0·8729	−0·4311	−0·5231	−1·2255	−1·2947
0·7887	0·2532	0·1939	0·1285	−1·0621	0·3646	−1·5349	1·8868	−0·6109	0·5067
0·9046	−0·9363	0·2668	−0·7122	0·9133	0·1885	0·3377	0·6994	−0·7522	0·5021
0·5823	0·1137	−1·1334	−0·9633	1·2725	1·2193	−1·5498	1·3002	0·4787	−0·0994
0·7836	0·4104	0·2755	0·6117	1·0492	0·2024	−1·5027	1·0811	−1·1240	1·0013
0·1600	−0·3707	0·3674	1·0033	−1·1969	0·4722	−1·6761	1·5762	−2·0142	0·8129
1·0383	−0·4590	−0·1122	2·1098	0·2146	−0·2230	0·1433	0·7726	−0·7707	−0·4194
0·1671	2·1522	0·2133	0·8366	−0·3594	−0·4389	0·7375	−0·8558	0·1095	−0·0173
0·3379	−2·0637	0·4198	−0·5608	0·7064	−1·8613	−0·5104	1·3333	2·7767	0·5664
0·3444	−0·0948	−1·8812	−2·6278	0·4124	0·8646	−0·6110	−0·1512	−1·8031	−0·7216
−0·3912	−0·7296	0·0778	0·2239	0·5802	−2·0438	−0·7473	−0·0416	2·1364	−0·3349
−0·5941	−0·5887	0·8105	−0·6841	0·0556	0·2533	0·4011	1·1621	0·1295	0·8688
−0·0768	0·2325	−1·2370	0·2177	2·0196	−0·4953	−0·2193	−0·2743	0·2454	1·4110
2·3696	−0·5056	−1·9071	0·8258	0·5697	0·6571	−0·0917	1·8961	2·1558	−1·4011
−0·3968	−0·5669	−0·3260	−0·6252	0·4869	0·0679	−0·5749	0·3863	−1·4732	0·6272
1·1401	0·7913	0·4862	0·9517	−1·3296	9·9970	−0·4263	−0·0943	2·1097	1·2943
−0·1992	0·9914	−0·8312	0·7272	−0·1765	0·6705	−0·1614	−0·7132	0·5529	0·6484
0·7894	1·7055	−1·9095	−0·2223	−0·2777	−0·1985	0·5114	0·2481	−0·1047	0·9714
0·7738	1·1404	−0·7707	−0·2558	−0·3052	−0·1825	1·2094	−0·0814	0·0982	−0·2936
−1·3150	0·1964	0·3990	−1·5290	−1·2785	0·7350	1·0500	1·5481	−0·6845	−1·1208
1·2719	−0·2309	0·9649	−0·7695	−0·3767	−1·6734	0·6314	0·2236	1·6357	0·7254
1·6399	0·9651	−1·5953	0·0153	1·0193	−1·8237	1·5742	−0·7431	1·2928	−1·1351
0·7164	−1·1403	−0·0382	−2·6190	−1·6919	1·4721	0·5274	−0·3755	0·1785	1·4302
1·9058	0·2461	−0·4089	0·3838	−0·1679	−0·3730	−0·8716	1·3133	0·6226	−0·0962
−0·5440	0·8169	0·7125	0·5219	−0·0041	0·1049	0·3602	0·4986	0·6017	−1·1846
−1·3378	0·0790	1·1813	0·3779	−0·3111	−0·1819	0·6715	−1·2309	1·1673	0·4598
1·5939	−0·7225	−0·8344	−1·9919	0·7127	1·2704	−1·0210	0·2453	−1·4375	0·5354
0·7378	0·9014	−0·8015	0·1428	−0·3749	0·6109	−0·6933	−1·1675	1·0786	0·6161
1·4320	0·0722	−0·6237	0·1263	−1·0313	0·5464	0·1833	−0·8393	0·6177	1·0176
0·5276	−0·7849	1·5668	0·1663	1·8116	−1·4139	−1·6417	−0·2523	−0·4680	−0·6125
−2·1245	−1·0242	0·4521	−0·2830	−1·9062	−0·4379	−0·3937	1·4846	0·5638	−1·7724
1·6638	−1·4929	0·6259	1·2061	2·2544	0·1115	−0·0851	0·8527	0·3650	0·5089
−0·0429	−1·4433	−0·3211	−1·4135	−0·6327	0·6226	−0·2088	0·5541	−0·1867	1·3761
−0·1320	−1·2630	0·7890	0·2889	1·6459	0·3946	−0·7841	−0·3490	−0·0772	1·1889
−0·2098	−0·7494	−1·2187	−0·1720	0·8910	−0·0909	−0·9212	−0·6525	0·9166	1·5315
−0·1563	−0·9592	−2·6399	0·7673	1·1387	−0·2521	−2·4283	−1·0642	1·4564	−0·7601
−0·8492	−1·1750	−0·7991	−1·9853	0·6764	−0·5518	0·0046	−1·0606	−0·2898	−0·2105
−1·8692	−1·2106	1·0306	0·2794	0·8007	−1·1584	0·4431	0·9379	1·0821	−0·6962

TABLE 29. BAYESIAN CONFIDENCE LIMITS FOR A BINOMIAL PARAMETER

If r is an observation from a binomial distribution (Table 1) of known index n and unknown parameter p, then, for an assigned probability C per cent, the pair of entries gives a C per cent Bayesian confidence interval for p. That is, there is C per cent probability that p lies between the values given. The intervals are the shortest possible, compatible with the requirement on probability. The tabulation is restricted to $r \le \frac{1}{2}n$. If $r > \frac{1}{2}n$, replace r by $n-r$ and take 1 minus the tabulated entries, in reverse order.

Example 1. $r = 7, n = 12$. Use $n = 12$ and $r = 5$ in the Table, which at a confidence level of 95 per cent gives 0·1856 and 0·6768, yielding the interval 0·3232 to 0·8144.

The intervals have been calculated using the reference prior which is uniform over the entire range $(0, 1)$ of p. The entries can be used for any beta prior with density proportional to $p^a(1-p)^b$, where a and b are non-negative integers, by replacing r with $r + a$ and n with $n + a + b$. If $r + a$ is outside the tabulated range, replace $r + a$ with $n - r + b$ and n with $n + a + b$, and take 1 minus the entries, in reverse order.

Example 2. $r = 7, n = 12$. If the prior has $a = 2$, $b = 1$, then $r + a = 9$, $n + a + b = 15$ and $n - r + b = 6$. Use $n = 15$

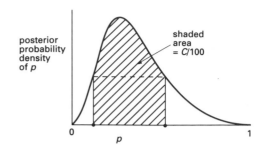

(This shape applies only when $0 < r < n$. When $r = 0$ or $r = n$, the intervals are one-sided.)

and $r = 6$ in the Table, which at a confidence level of 95 per cent gives 0·1909 and 0·6381, yielding the interval 0·3619 to 0·8091.

When n exceeds 30, C per cent limits for p are given approximately by

$$\hat{p} \pm x(P)[\hat{p}(1 - \hat{p})/n]^{\frac{1}{2}}$$

where $\hat{p} = r/n$, $P = \frac{1}{2}(100 - C)$ and $x(P)$ is the P percentage point of the normal distribution (Table 5).

CONFIDENCE LEVEL PER CENT

	90		95		99		99·9	
$n = 1$								
$r = 0$	0·0000	0·6838	0·0000	0·7764	0·0000	0·9000	0·0000	0·9684
$n = 2$								
$r = 0$	0·0000	0·5358	0·0000	0·6316	0·0000	0·7846	0·0000	0·9000
1	·1354	·8646	·0943	·9057	·0414	·9586	·0130	·9870
$n = 3$								
$r = 0$	0·0000	0·4377	0·0000	0·5271	0·0000	0·6838	0·0000	0·8222
1	·0679	·7122	·0438	·7723	·0159	·8668	·0037	·9377
$n = 4$								
$r = 0$	0·0000	0·3690	0·0000	0·4507	0·0000	0·6019	0·0000	0·7488
1	·0425	·6048	·0260	·6701	·0083	·7820	·0016	·8788
2	·1893	·8107	·1466	·8534	·0828	·9172	·0375	·9625
$n = 5$								
$r = 0$	0·0000	0·3187	0·0000	0·3930	0·0000	0·5358	0·0000	0·6838
1	·0302	·5253	·0178	·5906	·0052	·7083	·0009	·8186
2	·1380	·7111	·1048	·7613	·0567	·8441	·0242	·9133

TABLE 29. BAYESIAN CONFIDENCE LIMITS FOR A BINOMIAL PARAMETER

CONFIDENCE LEVEL PER CENT

	90		95		99		99·9	
n = 6								
r = 0	0·0000	0·2803	0·0000	0·3482	0·0000	0·4821	0·0000	0·6272
1	·0231	·4641	·0133	·5273	·0037	·6452	·0006	·7625
2	·1076	·6317	·0805	·6846	·0421	·7769	·0171	·8616
3	·2253	·7747	·1841	·8159	·1177	·8823	·0639	·9361
n = 7								
r = 0	0·0000	0·2505	0·0000	0·3123	0·0000	0·4377	0·0000	0·5783
1	·0185	·4155	·0105	·4759	·0028	·5913	·0004	·7113
2	·0878	·5677	·0650	·6210	·0331	·7174	·0129	·8115
3	·1839	·7008	·1488	·7459	·0934	·8227	·0495	·8912
n = 8								
r = 0	0·0000	0·2257	0·0000	0·2831	0·0000	0·4005	0·0000	0·5358
1	·0154	·3761	·0086	·4334	·0022	·5451	·0003	·6651
2	·0739	·5152	·0542	·5676	·0271	·6651	·0103	·7645
3	·1549	·6388	·1245	·6854	·0769	·7679	·0400	·8463
4	·2514	·7486	·2120	·7880	·1461	·8539	·0884	·9116
n = 9								
r = 0	0·0000	0·2057	0·0000	0·2589	0·0000	0·3690	0·0000	0·4988
1	·0132	·3435	·0073	·3978	·0018	·5053	·0003	·6237
2	·0638	·4714	·0464	·5224	·0229	·6192	·0085	·7212
3	·1337	·5863	·1068	·6332	·0652	·7184	·0333	·8032
4	·2165	·6901	·1816	·7316	·1237	·8039	·0739	·8714
n = 10								
r = 0	0·0000	0·1889	0·0000	0·2384	0·0000	0·3421	0·0000	0·4663
1	·0115	·3160	·0063	·3675	·0016	·4706	·0002	·5866
2	·0560	·4344	·0406	·4837	·0197	·5788	·0072	·6817
3	·1175	·5416	·0934	·5880	·0564	·6741	·0284	·7627
4	·1899	·6393	·1586	·6818	·1071	·7578	·0632	·8320
5	0·2712	0·7288	0·2338	0·7662	0·1693	0·8307	0·1100	0·8900
n = 11								
r = 0	0·0000	0·1746	0·0000	0·2209	0·0000	0·3187	0·0000	0·4377
1	·0102	·2926	·0055	·3415	·0013	·4402	·0002	·5534
2	·0499	·4027	·0360	·4502	·0173	·5431	·0062	·6456
3	·1047	·5030	·0829	·5485	·0497	·6344	·0248	·7252
4	·1691	·5951	·1407	·6377	·0943	·7156	·0551	·7943
5	0·2411	0·6803	0·2070	0·7191	0·1488	0·7878	0·0958	0·8539

See page 80 for explanation of the use of this table.

TABLE 29. BAYESIAN CONFIDENCE LIMITS FOR A BINOMIAL PARAMETER

CONFIDENCE LEVEL PER CENT

	90		95		99		99·9	
n = 12								
r = 0	0·0000	0·1623	0·0000	0·2058	0·0000	0·2983	0·0000	0·4122
1	·0091	·2724	·0049	·3188	·0012	·4134	·0002	·5234
2	·0449	·3753	·0323	·4210	·0154	·5113	·0055	·6128
3	·0944	·4695	·0745	·5138	·0443	·5987	·0219	·6905
4	·1524	·5564	·1263	·5987	·0841	·6773	·0488	·7588
5	0·2169	0·6374	0·1856	0·6768	0·1326	0·7479	0·0848	0·8188
6	·2870	·7130	·2513	·7487	·1887	·8113	·1290	·8710
n = 13								
r = 0	0·0000	0·1517	0·0000	0·1926	0·0000	0·2803	0·0000	0·3895
1	·0082	·2548	·0044	·2990	·0011	·3896	·0001	·4963
2	·0409	·3514	·0293	·3953	·0139	·4829	·0049	·5828
3	·0859	·4400	·0676	·4832	·0400	·5666	·0196	·6585
4	·1386	·5223	·1146	·5639	·0759	·6424	·0437	·7257
5	0·1971	0·5994	0·1682	0·6388	0·1195	0·7112	0·0759	0·7855
6	·2604	·6717	·2274	·7082	·1698	·7738	·1154	·8384
n = 14								
r = 0	0·0000	0·1423	0·0000	0·1810	0·0000	0·2644	0·0000	0·3691
1	·0075	·2394	·0040	·2814	·0009	·3684	·0001	·4718
2	·0375	·3303	·0267	·3726	·0126	·4574	·0044	·5554
3	·0788	·4140	·0619	·4559	·0364	·5376	·0177	·6290
4	·1271	·4921	·1048	·5329	·0691	·6106	·0396	·6948
5	0·1806	0·5654	0·1537	0·6045	0·1087	0·6775	0·0687	0·7539
6	·2383	·6346	·2075	·6715	·1542	·7388	·1043	·8070
7	·3000	·7000	·2659	·7341	·2051	·7949	·1457	·8543
n = 15								
r = 0	0·0000	0·1340	0·0000	0·1708	0·0000	0·2505	0·0000	0·3506
1	·0069	·2257	·0037	·2658	·0009	·3493	·0001	·4495
2	·0346	·3116	·0246	·3522	·0115	·4344	·0040	·5303
3	·0728	·3909	·0570	·4315	·0334	·5113	·0162	·6017
4	·1173	·4650	·0966	·5049	·0634	·5817	·0361	·6660
5	0·1666	0·5349	0·1415	0·5736	0·0997	0·6465	0·0627	0·7242
6	·2197	·6012	·1909	·6381	·1413	·7063	·0951	·7770
7	·2762	·6641	·2442	·6988	·1876	·7615	·1327	·8246

See page 80 for explanation of the use of this table.

TABLE 29. BAYESIAN CONFIDENCE LIMITS FOR A BINOMIAL PARAMETER

CONFIDENCE LEVEL PER CENT

	90		95		99		99·9	
n = 16								
r = 0	0·0000	0·1267	0·0000	0·1616	0·0000	0·2373	0·0000	0·3339
1	·0064	·2135	·0034	·2518	·0008	·3320	·0001	·4292
2	·0321	·2949	·0228	·3340	·0106	·4135	·0037	·5073
3	·0676	·3703	·0528	·4095	·0308	·4874	·0148	·5766
4	·1090	·4408	·0895	·4797	·0585	·5552	·0332	·6393
5	0·1546	0·5075	0·1311	0·5455	0·0920	0·6180	0·0576	0·6964
6	·2037	·5710	·1767	·6076	·1303	·6762	·0873	·7486
7	·2558	·6314	·2258	·6664	·1728	·7304	·1217	·7962
8	·3108	·6892	·2781	·7219	·2193	·7807	·1606	·8394
n = 17								
r = 0	0·0000	0·1201	0·0000	0·1533	0·0000	0·2257	0·0000	0·3187
1	·0060	·2025	·0032	·2393	·0007	·3164	·0001	·4105
2	·0300	·2799	·0213	·3175	·0099	·3945	·0034	·4860
3	·0631	·3516	·0492	·3897	·0286	·4656	·0137	·5533
4	·1017	·4189	·0834	·4568	·0544	·5310	·0307	·6144
5	0·1442	0·4827	0·1221	0·5200	0·0854	0·5917	0·0533	0·6703
6	·1899	·5435	·1644	·5798	·1209	·6483	·0807	·7217
7	·2383	·6017	·2099	·6366	·1601	·7013	·1124	·7690
8	·2893	·6574	·2583	·6905	·2030	·7508	·1481	·8124
n = 18								
r = 0	0·0000	0·1141	0·0000	0·1459	0·0000	0·2152	0·0000	0·3048
1	·0056	·1926	·0030	·2279	·0007	·3021	·0001	·3934
2	·0281	·2663	·0199	·3026	·0092	·3771	·0031	·4665
3	·0592	·3348	·0461	·3716	·0267	·4455	·0128	·5317
4	·0953	·3991	·0781	·4360	·0508	·5086	·0286	·5912
5	0·1351	0·4602	0·1142	0·4967	0·0797	0·5674	0·0496	0·6459
6	·1778	·5185	·1537	·5543	·1127	·6224	·0750	·6964
7	·2230	·5745	·1962	·6092	·1492	·6741	·1044	·7432
8	·2705	·6282	·2412	·6615	·1889	·7228	·1374	·7864
9	·3201	·6799	·2886	·7114	·2316	·7684	·1738	·8262
n = 19								
r = 0	0·0000	0·1087	0·0000	0·1391	0·0000	0·2057	0·0000	0·2920
1	·0052	·1836	·0028	·2175	·0006	·2891	·0001	·3776
2	·0264	·2540	·0187	·2890	·0086	·3612	·0029	·4484
3	·0557	·3194	·0433	·3551	·0251	·4271	·0119	·5117
4	·0897	·3810	·0734	·4170	·0476	·4880	·0267	·5696
5	0·1271	0·4396	0·1073	0·4754	0·0747	0·5449	0·0463	0·6231
6	·1672	·4957	·1444	·5309	·1056	·5984	·0701	·6726
7	·2096	·5495	·1841	·5839	·1397	·6488	·0974	·7187
8	·2540	·6014	·2261	·6346	·1766	·6964	·1281	·7615
9	·3004	·6514	·2704	·6832	·2164	·7413	·1619	·8013

See page 80 for explanation of the use of this table.

TABLE 29. BAYESIAN CONFIDENCE LIMITS FOR A BINOMIAL PARAMETER

	90		95		99		99·9	
n = 20								
r = 0	0·0000	0·1039	0·0000	0·1329	0·0000	0·1969	0·0000	0·2803
1	·0049	·1754	·0026	·2080	·0006	·2771	·0001	·3630
2	·0249	·2428	·0176	·2766	·0081	·3466	·0027	·4315
3	·0526	·3055	·0409	·3401	·0236	·4101	·0112	·4931
4	·0847	·3645	·0692	·3995	·0448	·4690	·0251	·5494
5	0·1200	0·4208	0·1012	0·4557	0·0703	0·5241	0·0435	0·6016
6	·1578	·4747	·1361	·5093	·0993	·5760	·0657	·6501
7	·1977	·5266	·1734	·5606	·1312	·6253	·0913	·6955
8	·2395	·5767	·2129	·6097	·1659	·6717	·1199	·7379
9	·2828	·6253	·2544	·6569	·2030	·7158	·1514	·7775
10	0·3281	0·6719	0·2978	0·7022	0·2425	0·7575	0·1856	0·8144
n = 21								
r = 0	0·0000	0·0994	0·0000	0·1273	0·0000	0·1889	0·0000	0·2695
1	·0047	·1679	·0025	·1994	·0006	·2661	·0001	·3494
2	·0236	·2325	·0167	·2652	·0076	·3331	·0026	·4159
3	·0498	·2926	·0387	·3262	·0223	·3944	·0105	·4757
4	·0802	·3494	·0655	·3834	·0423	·4514	·0236	·5306
5	0·1136	0·4035	0·0957	0·4376	0·0664	0·5048	0·0409	0·5815
6	·1493	·4554	·1287	·4894	·0937	·5552	·0619	·6290
7	·1870	·5055	·1639	·5389	·1238	·6030	·0859	·6735
8	·2265	·5539	·2011	·5866	·1563	·6485	·1128	·7153
9	·2675	·6007	·2402	·6324	·1912	·6917	·1423	·7546
10	0·3099	0·6461	0·2809	0·6766	0·2281	0·7328	0·1742	0·7914
n = 22								
r = 0	0·0000	0·0953	0·0000	0·1221	0·0000	0·1815	0·0000	0·2594
1	·0044	·1611	·0023	·1914	·0005	·2557	·0001	·3369
2	·0224	·2231	·0158	·2547	·0072	·3206	·0024	·4014
3	·0473	·2809	·0367	·3134	·0211	·3798	·0099	·4594
4	·0762	·3354	·0621	·3686	·0401	·4350	·0223	·5129
5	0·1079	0·3875	0·0908	0·4209	0·0628	0·4868	0·0387	0·5626
6	·1418	·4376	·1220	·4709	·0887	·5358	·0584	·6091
7	·1775	·4859	·1554	·5189	·1171	·5823	·0811	·6528
8	·2148	·5327	·1905	·5651	·1478	·6267	·1064	·6939
9	·2536	·5781	·2274	·6096	·1807	·6690	·1341	·7328
10	0·2937	0·6221	0·2659	0·6526	0·2154	0·7094	0·1641	0·7693
11	·3351	·6649	·3059	·6941	·2521	·7479	·1963	·8037

See page 80 for explanation of the use of this table.

TABLE 29. BAYESIAN CONFIDENCE LIMITS FOR A BINOMIAL PARAMETER

	90		95		99		99·9	
n = 23								
r = 0	0·0000	0·0915	0·0000	0·1173	0·0000	0·1746	0·0000	0·2505
1	·0042	·1547	·0022	·1840	·0005	·2465	·0001	·3252
2	·0214	·2144	·0150	·2450	·0069	·3089	·0023	·3878
3	·0451	·2700	·0349	·3016	·0200	·3663	·0094	·4442
4	·0726	·3225	·0591	·3548	·0380	·4197	·0211	·4963
5	0·1027	0·3728	0·0864	0·4053	0·0597	0·4700	0·0367	0·5448
6	·1349	·4211	·1160	·4537	·0842	·5176	·0554	·5903
7	·1689	·4678	·1475	·5005	·1111	·5629	·0768	·6331
8	·2043	·5131	·1810	·5450	·1402	·6062	·1007	·6736
9	·2411	·5570	·2160	·5883	·1712	·6476	·1269	·7119
10	0·2791	0·5998	0·2524	0·6302	0·2041	0·6872	0·1551	0·7481
11	·3183	·6413	·2902	·6707	·2386	·7251	·1854	·7824
n = 24								
r = 0	0·0000	0·0880	0·0000	0·1129	0·0000	0·1682	0·0000	0·2414
1	·0040	·1489	·0021	·1772	·0005	·2377	·0001	·3142
2	·0204	·2063	·0143	·2360	·0065	·2981	·0022	·3753
3	·0430	·2599	·0333	·2906	·0191	·3537	·0090	·4300
4	·0692	·3106	·0564	·3420	·0362	·4055	·0201	·4807
5	0·0980	0·3591	0·0824	0·3909	0·0568	0·4543	0·0348	0·5280
6	·1287	·4058	·1106	·4377	·0799	·5011	·0526	·5725
7	·1610	·4510	·1407	·4828	·1057	·5447	·0729	·6145
8	·1948	·4948	·1724	·5263	·1333	·5869	·0956	·6543
9	·2297	·5374	·2056	·5684	·1627	·6274	·1203	·6920
10	0·2659	0·5789	0·2402	0·6092	0·1938	0·6662	0·1471	0·7279
11	·3031	·6193	·2760	·6487	·2265	·7035	·1756	·7618
12	·3414	·6586	·3131	·6869	·2607	·7393	·2060	·7940
n = 25								
r = 0	0·0000	0·0848	0·0000	0·1088	0·0000	0·1623	0·0000	0·2333
1	·0038	·1435	·0020	·1708	·0004	·2295	·0001	·3040
2	·0195	·1988	·0137	·2276	·0062	·2880	·0021	·3631
3	·0411	·2505	·0318	·2804	·0182	·3419	·0085	·4166
4	·0662	·2995	·0539	·3301	·0346	·3921	·0191	·4660
5	0·0937	0·3464	0·0787	0·3775	0·0542	0·4395	0·0332	0·5122
6	·1230	·3916	·1056	·4228	·0764	·4846	·0501	·5557
7	·1539	·4353	·1344	·4665	·1008	·5276	·0694	·5969
8	·1861	·4778	·1646	·5088	·1271	·5688	·0910	·6360
9	·2194	·5191	·1962	·5497	·1551	·6084	·1145	·6731
10	0·2538	0·5594	0·2291	0·5894	0·1846	0·6464	0·1398	0·7085
11	·2893	·5986	·2632	·6279	·2156	·6830	·1668	·7421
12	·3257	·6369	·2983	·6654	·2480	·7182	·1955	·7741

See page 80 for explanation of the use of this table.

TABLE 29. BAYESIAN CONFIDENCE LIMITS FOR A BINOMIAL PARAMETER

CONFIDENCE LEVEL PER CENT

	90		95		99		99·9	
n = 26								
r = 0	0·0000	0·0817	0·0000	0·1050	0·0000	0·1568	0·0000	0·2257
1	·0037	·1384	·0019	·1649	·0004	·2219	·00005	·2944
2	·0187	·1919	·0131	·2198	·0059	·2786	·0020	·3519
3	·0394	·2418	·0305	·2709	·0174	·3308	·0081	·4040
4	·0634	·2892	·0516	·3190	·0331	·3796	·0183	·4522
5	0·0898	0·3345	0·0753	0·3649	0·0518	0·4257	0·0317	0·4973
6	·1179	·3783	·1011	·4089	·0731	·4696	·0478	·5399
7	·1474	·4206	·1286	·4513	·0963	·5115	·0663	·5802
8	·1781	·4618	·1575	·4924	·1214	·5517	·0868	·6185
9	·2100	·5019	·1877	·5322	·1481	·5904	·1091	·6551
10	0·2429	0·5411	0·2190	0·5708	0·1762	0·6276	0·1332	0·6899
11	·2767	·5793	·2515	·6084	·2057	·6635	·1589	·7232
12	·3114	·6166	·2850	·6450	·2365	·6981	·1861	·7549
13	·3470	·6530	·3195	·6805	·2686	·7314	·2148	·7852
n = 27								
r = 0	0·0000	0·0789	0·0000	0·1015	0·0000	0·1517	0·0000	0·2186
1	·0035	·1337	·0018	·1594	·0004	·2148	·00005	·2854
2	·0179	·1854	·0126	·2125	·0057	·2698	·0019	·3414
3	·0378	·2337	·0292	·2620	·0167	·3205	·0078	·3921
4	·0609	·2796	·0495	·3086	·0317	·3679	·0175	·4391
5	0·0861	0·3235	0·0723	0·3531	0·0496	0·4127	0·0303	0·4832
6	·1131	·3658	·0970	·3958	·0700	·4554	·0457	·5248
7	·1414	·4069	·1233	·4370	·0922	·4963	·0634	·5643
8	·1708	·4469	·1509	·4770	·1162	·5355	·0829	·6019
9	·2014	·4859	·1798	·5157	·1417	·5733	·1043	·6379
10	0·2328	0·5239	0·2098	0·5534	0·1685	0·6098	0·1272	0·6722
11	·2651	·5611	·2408	·5900	·1966	·6450	·1516	·7051
12	·2983	·5974	·2728	·6257	·2260	·6790	·1775	·7365
13	·3323	·6330	·3057	·6605	·2565	·7118	·2048	·7665
n = 28								
r = 0	0·0000	0·0763	0·0000	0·0981	0·0000	0·1468	0·0000	0·2119
1	·0034	·1293	·0018	·1543	·0004	·2081	·00004	·2769
2	·0172	·1793	·0121	·2057	·0055	·2615	·0018	·3314
3	·0364	·2261	·0281	·2537	·0160	·3107	·0075	·3809
4	·0585	·2705	·0475	·2989	·0304	·3569	·0168	·4268
5	0·0828	0·3131	0·0694	0·3421	0·0476	0·4005	0·0290	0·4698
6	·1087	·3542	·0931	·3836	·0672	·4420	·0438	·5106
7	·1358	·3941	·1184	·4236	·0885	·4819	·0607	·5493
8	·1641	·4329	·1449	·4624	·1114	·5203	·0794	·5862
9	·1934	·4708	·1724	·5004	·1358	·5572	·0998	·6215

See page 80 for explanation of the use of this table.

TABLE 29. BAYESIAN CONFIDENCE LIMITS FOR A BINOMIAL PARAMETER

	90		95		99		99·9	
n = 28								
r = 10	0·2235	0·5078	0·2013	0·5369	0·1615	0·5929	0·1217	0·6553
11	·2545	·5439	·2310	·5727	·1884	·6274	·1450	·6877
12	·2863	·5794	·2615	·6075	·2164	·6608	·1697	·7188
13	·3188	·6140	·2930	·6415	·2455	·6931	·1957	·7486
14	·3520	·6480	·3253	·6747	·2757	·7243	·2229	·7771
n = 29								
r = 0	0·0000	0·0739	0·0000	0·0950	0·0000	0·1423	0·0000	0·2057
1	·0032	·1253	·0017	·1494	·0004	·2018	·00004	·2689
2	·0166	·1737	·0116	·1993	·0052	·2537	·0017	·3220
3	·0350	·2190	·0270	·2458	·0154	·3016	·0072	·3703
4	·0564	·2621	·0458	·2898	·0292	·3465	·0161	·4151
5	0·0797	0·3034	0·0668	0·3317	0·0458	0·3890	0·0279	0·4572
6	·1046	·3433	·0896	·3720	·0645	·4295	·0421	·4970
7	·1307	·3820	·1138	·4110	·0850	·4684	·0582	·5349
8	·1579	·4197	·1393	·4488	·1070	·5058	·0762	·5711
9	·1860	·4566	·1659	·4855	·1304	·5419	·0957	·6058
10	0·2150	0·4926	0·1935	0·5213	0·1550	0·5769	0·1167	0·6391
11	·2447	·5278	·2219	·5562	·1807	·6107	·1390	·6710
12	·2752	·5623	·2512	·5903	·2075	·6435	·1626	·7017
13	·3063	·5962	·2814	·6236	·2354	·6752	·1873	·7312
14	·3382	·6293	·3123	·6560	·2642	·7060	·2133	·7595
n = 30								
r = 0	0·0000	0·0716	0·0000	0·0921	0·0000	0·1380	0·0000	0·1997
1	·0031	·1213	·0016	·1449	·0004	·1958	·00004	·2613
2	·0160	·1683	·0112	·1933	·0050	·2463	·0017	·3131
3	·0338	·2123	·0260	·2385	·0148	·2929	·0069	·3602
4	·0543	·2541	·0441	·2812	·0281	·3366	·0155	·4040
5	0·0768	0·2942	0·0644	0·3219	0·0441	0·3780	0·0268	0·4452
6	·1008	·3330	·0863	·3612	·0621	·4176	·0404	·4842
7	·1260	·3707	·1097	·3991	·0818	·4555	·0560	·5213
8	·1522	·4073	·1342	·4359	·1030	·4921	·0732	·5568
9	·1792	·4432	·1597	·4717	·1254	·5274	·0919	·5909
10	0·2071	0·4782	0·1862	0·5066	0·1490	0·5616	0·1120	0·6236
11	·2357	·5126	·2136	·5407	·1737	·5948	·1334	·6551
12	·2649	·5462	·2417	·5740	·1994	·6270	·1560	·6854
13	·2949	·5793	·2706	·6065	·2261	·6582	·1797	·7146
14	·3254	·6116	·3002	·6383	·2536	·6885	·2045	·7426
15	0·3566	0·6434	0·3306	0·6694	0·2821	0·7179	0·2304	0·7696

See page 80 for explanation of the use of this table.

TABLE 30. BAYESIAN CONFIDENCE LIMITS FOR A POISSON MEAN

If $x_1, x_2, \ldots, x_n$ is a random sample of size n from a Poisson distribution (Table 2) of unknown mean μ, and $r = \sum_{i=1}^{n} x_i$, then, for an assigned probability C per cent, the pair of entries *when divided by* n gives a C per cent Bayesian confidence interval for μ. That is, there is C per cent probability that μ lies between the values given. The intervals are the shortest possible, compatible with the requirement on probability.

Example. $r = 30$, $n = 10$. With a confidence level of 95 per cent, the Table at $r = 30$ gives 19·66 and 40·91. On division by $n = 10$, the required interval is 1·966 to 4·091. The intervals have been calculated using the reference prior with density proportional to μ^{-1}, and the posterior density is such that $n\mu = \frac{1}{2}\chi^2_{2r}$ (Table 8). The entries can be used for any gamma prior with density

$$\exp(-m\mu)\mu^{s-1}m^s/(s-1)!,$$

where m and s are non-negative integers, by replacing n with $m + n$ and r with $r + s$. No limits are available in the extreme case $r = 0$.

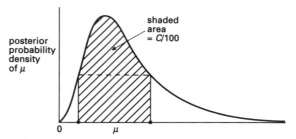

(This shape applies only when $r \geq 2$. When $r = 1$, the intervals are one-sided.)

When r exceeds 45, C per cent limits for μ are given approximately by

$$\frac{r}{n} \pm x(P)\frac{r^{\frac{1}{2}}}{n}$$

where $P = \frac{1}{2}(100 - C)$ and $x(P)$ is the P percentage point of the normal distribution (Table 5).

CONFIDENCE LEVEL PER CENT

	90		95		99		99·9	
$r = 1$	0·000	2·303	0·000	2·996	0·000	4·605	0·000	6·908
2	0·084	3·932	0·042	4·765	0·009	6·638	0·001	9·233
3	0·441	5·479	0·304	6·401	0·132	8·451	0·042	11·24
4	0·937	6·946	0·713	7·948	0·393	10·15	0·176	13·11
5	1·509	8·355	1·207	9·430	0·749	11·77	0·399	14·88
6	2·129	9·723	1·758	10·86	1·172	13·33	0·691	16·58
7	2·785	11·06	2·350	12·26	1·646	14·84	1·040	18·22
8	3·467	12·37	2·974	13·63	2·158	16·32	1·433	19·83
9	4·171	13·66	3·623	14·98	2·702	17·77	1·862	21·39
10	4·893	14·94	4·292	16·30	3·272	19·19	2·323	22·93
11	5·629	16·20	4·979	17·61	3·864	20·60	2·811	24·44
12	6·378	17·45	5·681	18·91	4·476	21·98	3·321	25·92
13	7·138	18·69	6·395	20·19	5·104	23·35	3·852	27·39
14	7·908	19·91	7·122	21·46	5·746	24·71	4·401	28·84
15	8·686	21·14	7·858	22·73	6·402	26·05	4·965	30·27
16	9·472	22·35	8·603	23·98	7·069	27·38	5·545	31·69
17	10·26	23·55	9·355	25·23	7·747	28·70	6·137	33·10
18	11·06	24·75	10·12	26·46	8·434	30·01	6·741	34·50
19	11·87	25·95	10·89	27·69	9·131	31·32	7·356	35·88
20	12·68	27·14	11·66	28·92	9·835	32·61	7·981	37·25
21	13·49	28·32	12·44	30·14	10·55	33·90	8·616	38·62
22	14·31	29·50	13·22	31·35	11·27	35·18	9·259	39·97
23	15·14	30·68	14·01	32·56	11·99	36·45	9·910	41·32
24	15·96	31·85	14·81	33·77	12·72	37·72	10·57	42·66

TABLE 30. BAYESIAN CONFIDENCE LIMITS FOR A POISSON MEAN

CONFIDENCE LEVEL PER CENT

	90		95		99		99·9	
$r = 25$	16·80	33·02	15·61	34·97	13·46	38·98	11·24	44·00
26	17·63	34·18	16·41	36·16	14·20	40·24	11·91	45·32
27	18·47	35·35	17·22	37·35	14·95	41·49	12·59	46·64
28	19·31	36·50	18·03	38·54	15·70	42·74	13·27	47·96
29	20·15	37·66	18·84	39·73	16·46	43·98	13·96	49·27
30	21·00	38·81	19·66	40·91	17·22	45·22	14·66	50·57
31	21·85	39·96	20·48	42·09	17·98	46·45	15·36	51·87
32	22·70	41·11	21·31	43·27	18·75	47·68	16·06	53·16
33	23·55	42·26	22·13	44·44	19·52	48·91	16·78	54·45
34	24·41	43·40	22·96	45·61	20·30	50·14	17·49	55·74
35	25·27	44·54	23·79	46·78	21·08	51·36	18·21	57·02
36	26·13	45·68	24·63	47·94	21·86	52·57	18·93	58·30
37	26·99	46·82	25·46	49·11	22·65	53·79	19·66	59·57
38	27·86	47·95	26·30	50·27	23·43	55·00	20·39	60·84
39	28·72	49·09	27·14	51·43	24·23	56·21	21·12	62·10
40	29·59	50·22	27·98	52·58	25·02	57·41	21·86	63·37
41	30·46	51·35	28·83	53·74	25·82	58·62	22·60	64·63
42	31·33	52·48	29·68	54·89	26·62	59·82	23·35	65·88
43	32·20	53·60	30·52	56·04	27·42	61·02	24·09	67·13
44	33·08	54·73	31·37	57·19	28·22	62·21	24·84	68·38
45	33·95	55·85	32·23	58·34	29·03	63·41	25·59	69·63

TABLE 31. BAYESIAN CONFIDENCE LIMITS FOR THE SQUARE OF A MULTIPLE CORRELATION COEFFICIENT

For a normal distribution of $(k + 1)$ quantities, let ρ^2 be the square of the true multiple correlation coefficient between the first quantity and the remaining k (sometimes called 'explanatory variables'). ρ^2 is the proportion of the variance of the first quantity that is accounted for by the remaining k. If R^2 denotes the square of the corresponding sample multiple correlation coefficient from a random sample of size n ($n > k + 1$), then, for an assigned probability C per cent, the pair of entries gives a C per cent Bayesian confidence interval for ρ^2. That is, there is C per cent probability that ρ^2 lies between the values given. The entries have been calculated using a reference prior which is uniform over the entire range $(0,1)$ of ρ^2. The intervals are the shortest possible, compatible with the requirement on probability. When $R^2 = 1$, both the upper and lower limits may be taken to be 1.

Interpolation in n and R^2 will often be needed. When n is large, C per cent limits for ρ^2 are given approximately by

$$R^2 \pm 2x(P)(1 - R^2)(R^2/n)^{\frac{1}{2}}$$

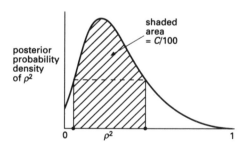

(In some cases this shape does not apply, and the intervals are one-sided.)

where $P = \frac{1}{2}(100 - C)$ and $x(P)$ is the P percentage point of the normal distribution ('Table 5). More accurate upper limits are found by harmonic interpolation (see page 96) in the function $f(n) = \sqrt{n}(U(n) - R^2)$, where $U(n)$ is the upper limit for sample size n. For the lower limit, $L(n)$, use the function $f(n) = \sqrt{n}(R^2 - L(n))$; in each case $f(\infty) = 2x(P)(1 - R^2)\sqrt{R^2}$.

TABLE 31. BAYESIAN CONFIDENCE LIMITS FOR THE SQUARE OF A MULTIPLE CORRELATION COEFFICIENT

$k = 1$

CONFIDENCE LEVEL PER CENT

	90		95		99		99·9	
$n = 3$								
$R^2 = 0\cdot00$	0·0000	0·6838	0·0000	0·7764	0·0000	0·9000	0·0000	0·9683
·10	·0000	·6985	·0000	·7882	·0000	·9060	·0000	·9704
·20	·0000	·7140	·0000	·8004	·0000	·9122	·0000	·9725
·30	·0000	·7305	·0000	·8132	·0000	·9186	·0000	·9746
·40	·0000	·7482	·0000	·8269	·0000	·9253	·0000	·9768
0·50	0·0000	0·7676	0·0000	0·8416	0·0000	0·9325	0·0000	0·9793
·60	·0000	·7892	·0000	·8579	·0000	·9402	·0000	·9817
·70	·0000	·8142	·0000	·8764	·0000	·9489	·0000	·9845
·80	·0391	·8810	·0000	·8987	·0000	·9592	·0000	·9878
·90	·1155	·9644	·0512	·9672	·0000	·9724	·0000	·9919
0·95	0·1525	0·9887	0·0767	0·9898	0·0118	0·9905	0·0000	0·9948
$n = 10$								
$R^2 = 0\cdot00$	0·0000	0·3421	0·0000	0·4200	0·0000	0·5671	0·0000	0·7152
·10	·0000	·4481	·0000	·5264	·0000	·6621	·0000	·7864
·20	·0000	·5202	·0000	·5938	·0000	·7163	·0000	·8238
·30	·0000	·5809	·0000	·6491	·0000	·7591	·0000	·8525
·40	·0538	·6754	·0170	·7139	·0000	·7968	·0000	·8772
0·50	0·1167	0·7518	0·0666	0·7864	0·0029	0·8347	0·0000	0·8996
·60	·1950	·8183	·1310	·8466	·0384	·8877	·0000	·9206
·70	·2959	·8770	·2189	·8982	·0931	·9287	·0124	·9491
·80	·4328	·9275	·3471	·9413	·1867	·9608	·0531	·9744
·90	·6357	·9690	·5563	·9756	·3781	·9847	·1720	·9909
0·95	0·7846	0·9860	0·7259	0·9892	0·5735	0·9936	0·3448	0·9964
$n = 25$								
$R^2 = 0\cdot00$	0·0000	0·1623	0·0000	0·2058	0·0000	0·2983	0·0000	0·4122
·10	·0000	·3128	·0000	·3660	·0000	·4665	·0000	·5747
·20	·0262	·4230	·0090	·4655	·0000	·5536	·0000	·6512
·30	·0817	·5222	·0530	·5622	·0129	·6339	·0000	·7116
·40	·1544	·6106	·1158	·6464	·0535	·7091	·0090	·7696
0·50	0·2439	0·6906	0·1980	0·7215	0·1166	0·7746	0·0445	0·8248
·60	·3507	·7637	·3009	·7890	·2058	·8320	·1081	·8719
·70	·4762	·8306	·4270	·8500	·3266	·8824	·2101	·9121
·80	·6232	·8921	·5807	·9052	·4881	·9268	·3678	·9463
·90	·7958	·9485	·7683	·9551	·7046	·9659	·6116	·9754
0·95	0·8935	0·9749	0·8778	0·9782	0·8403	0·9836	0·7821	0·9883

See page 89 for explanation of the use of this table.

TABLE 31. BAYESIAN CONFIDENCE LIMITS FOR THE SQUARE OF A MULTIPLE CORRELATION COEFFICIENT

$$k = 2$$

CONFIDENCE LEVEL PER CENT

	90		95		99		99·9	
n = 4								
$R^2 = 0\cdot00$	0·0000	0·6019	0·0000	0·6983	0·0000	0·8415	0·0000	0·9369
·10	·0000	·6179	·0000	·7123	·0000	·8503	·0000	·9408
·20	·0000	·6352	·0000	·7272	·0000	·8595	·0000	·9448
·30	·0000	·6541	·0000	·7434	·0000	·8694	·0000	·9491
·40	·0000	·6751	·0000	·7611	·0000	·8799	·0000	·9535
0·50	0·0000	0·6986	0·0000	0·7807	0·0000	0·8914	0·0000	0·9586
·60	·0000	·7255	·0000	·8028	·0000	·9040	·0000	·9636
·70	·0000	·7573	·0000	·8284	·0000	·9184	·0000	·9695
·80	·0000	·7969	·0000	·8597	·0000	·9353	·0000	·9765
·90	·0614	·9055	·0085	·9100	·0000	·9572	·0000	·9848
0·95	0·1189	0·9690	0·0540	0·9707	0·0000	0·9720	0·0000	0·9905
n = 10								
$R^2 = 0\cdot00$	0·0000	0·3421	0·0000	0·4200	0·0000	0·5671	0·0000	0·7152
·10	·0000	·4101	·0000	·4904	·0000	·6331	·0000	·7665
·20	·0000	·4741	·0000	·5527	·0000	·6857	·0000	·8040
·30	·0000	·5354	·0000	·6098	·0000	·7313	·0000	·8352
·40	·0000	·5953	·0000	·6641	·0000	·7728	·0000	·8626
0·50	0·0363	0·6854	0·0000	0·7170	0·0000	0·8118	0·0000	0·8877
·60	·1122	·7800	·0557	·8073	·0000	·8493	·0000	·9112
·70	·2107	·8557	·1351	·8779	·0307	·9062	·0000	·9338
·80	·3487	·9171	·2564	·9321	·1014	·9523	·0105	·9629
·90	·5639	·9656	·4692	·9728	·2688	·9825	·0774	·9887
0·95	0·7323	0·9848	0·6562	0·9883	0·4649	0·9929	0·2109	0·9959
n = 25								
$R^2 = 0\cdot00$	0·0000	0·1623	0·0000	0·2058	0·0000	0·2983	0·0000	0·4122
·10	·0000	·2822	·0000	·3364	·0000	·4400	·0000	·5525
·20	·0000	·3783	·0000	·4329	·0000	·5321	·0000	·6341
·30	·0503	·4938	·0242	·5334	·0000	·6079	·0000	·6983
·40	·1225	·5904	·0849	·6273	·0272	·6907	·0000	·7532
0·50	0·2130	0·6758	0·1667	0·7079	0·0866	0·7628	0·0211	0·8134
·60	·3221	·7530	·2708	·7794	·1742	·8241	·0786	·8653
·70	·4514	·8234	·4000	·8436	·2958	·8774	·1772	·9081
·80	·6040	·8878	·5589	·9015	·4611	·9240	·3352	·9442
·90	·7845	·9465	·7550	·9535	·6865	·9647	·5866	·9746
0·95	0·8873	0·9739	0·8705	0·9774	0·8298	0·9830	0·7664	0·9879

See page 89 for explanation of the use of this table.

TABLE 31. BAYESIAN CONFIDENCE LIMITS FOR THE SQUARE OF A MULTIPLE CORRELATION COEFFICIENT

$k = 3$

CONFIDENCE LEVEL PER CENT

	90		95		99		99·9	
$n = 5$								
$R^2 = 0\cdot00$	0·0000	0·5358	0·0000	0·6316	0·0000	0·7846	0·0000	0·9000
·10	·0000	·5532	·0000	·6477	·0000	·7962	·0000	·9061
·20	·0000	·5721	·0000	·6652	·0000	·8085	·0000	·9126
·30	·0000	·5931	·0000	·6842	·0000	·8217	·0000	·9196
·40	·0000	·6166	·0000	·7053	·0000	·8360	·0000	·9266
0·50	0·0000	0·6433	0·0000	0·7289	0·0000	0·8517	0·0000	0·9344
·60	·0000	·6743	·0000	·7558	·0000	·8691	·0000	·9430
·70	·0000	·7114	·0000	·7874	·0000	·8890	·0000	·9529
·80	·0000	·7582	·0000	·8262	·0000	·9125	·0000	·9634
·90	·0101	·8337	·0000	·8787	·0000	·9426	·0000	·9770
0·95	0·0951	0·9446	0·0362	0·9470	0·0000	0·9630	0·0000	0·9858
$n = 15$								
$R^2 = 0\cdot00$	0·0000	0·2505	0·0000	0·3123	0·0000	0·4377	0·0000	0·5784
·10	·0000	·3204	·0000	·3892	·0000	·5186	·0000	·6522
·20	·0000	·3930	·0000	·4630	·0000	·5879	·0000	·7093
·30	·0000	·4662	·0000	·5340	·0000	·6498	·0000	·7576
·40	·0000	·5395	·0000	·6026	·0000	·7070	·0000	·8002
0·50	0·0648	0·6565	0·0231	0·6885	0·0000	0·7602	0·0000	0·8388
·60	·1562	·7522	·0973	·7820	·0153	·8230	·0000	·8745
·70	·2779	·8309	·2053	·8548	·0823	·8901	·0045	·9117
·80	·4409	·8976	·3635	·9138	·2093	·9381	·0628	·9561
·90	·6656	·9539	·6021	·9620	·4537	·9739	·2546	·9830
0·95	0·8139	0·9783	0·7720	0·9823	0·6638	0·9882	0·4879	0·9926
$n = 25$								
$R^2 = 0\cdot00$	0·0000	0·1623	0·0000	0·2058	0·0000	0·2983	0·0000	0·4122
·10	·0000	·2580	·0000	·3123	·0000	·4175	·0000	·5330
·20	·0000	·3514	·0000	·4075	·0000	·5103	·0000	·6166
·30	·0190	·4556	·0000	·4941	·0000	·5895	·0000	·6841
·40	·0890	·5667	·0533	·6038	·0033	·6631	·0000	·7421
0·50	0·1797	0·6592	0·1333	0·6924	0·0562	0·7484	0·0014	0·7947
·60	·2909	·7412	·2382	·7688	·1407	·8152	·0491	·8572
·70	·4242	·8155	·3704	·8367	·2623	·8719	·1421	·9037
·80	·5827	·8830	·5349	·8974	·4314	·9209	·2995	·9419
·90	·7719	·9444	·7402	·9517	·6664	·9634	·5586	·9737
0·95	0·8804	0·9729	0·8621	0·9766	0·8179	0·9824	0·7484	0·9875

See page 89 for explanation of the use of this table.

TABLE 31. BAYESIAN CONFIDENCE LIMITS FOR THE SQUARE OF A MULTIPLE CORRELATION COEFFICIENT

$k = 4$

CONFIDENCE LEVEL PER CENT

	90		95		99		99·9	
$n = 6$								
$R^2 = 0·00$	0·0000	0·4820	0·0000	0·5751	0·0000	0·7317	0·0000	0·8610
·10	·0000	·5011	·0000	·5928	·0000	·7458	·0000	·8696
·20	·0000	·5203	·0000	·6121	·0000	·7609	·0000	·8786
·30	·0000	·5427	·0000	·6334	·0000	·7771	·0000	·8880
·40	·0000	·5681	·0000	·6571	·0000	·7948	·0000	·8982
0·50	0·0000	0·5971	0·0000	0·6839	0·0000	0·8143	0·0000	0·9092
·60	·0000	·6312	·0000	·7147	·0000	·8360	·0000	·9212
·70	·0000	·6724	·0000	·7512	·0000	·8610	·0000	·9346
·80	·0000	·7249	·0000	·7964	·0000	·8907	·0000	·9499
·90	·0000	·7993	·0000	·8581	·0000	·9287	·0000	·9689
0·95	0·0731	0·9167	0·0186	0·9199	0·0000	0·9543	0·0000	0·9810
$n = 15$								
$R^2 = 0·00$	0·0000	0·2505	0·0000	0·3123	0·0000	0·4377	0·0000	0·5784
·10	·0000	·3050	·0000	·3732	·0000	·5032	·0000	·6394
·20	·0000	·3669	·0000	·4377	·0000	·5661	·0000	·6927
·30	·0000	·4343	·0000	·5047	·0000	·6264	·0000	·7408
·40	·0000	·5061	·0000	·5731	·0000	·6846	·0000	·7849
0·50	0·0110	0·5916	0·0000	0·6424	0·0000	0·7406	0·0000	0·8258
·60	·0959	·7160	·0442	·7438	·0000	·7947	·0000	·8641
·70	·2130	·8117	·1409	·8363	·0347	·8689	·0000	·9001
·80	·3792	·8882	·2952	·9055	·1384	·9306	·0218	·9461
·90	·6200	·9505	·5463	·9592	·3770	·9719	·1662	·9813
0·95	0·7850	0·9769	0·7345	0·9812	0·6035	0·9874	0·3957	0·9921
$n = 25$								
$R^2 = 0·00$	0·0000	0·1623	0·0000	0·2058	0·0000	0·2983	0·0000	0·4122
·10	·0000	·2399	·0000	·2937	·0000	·3993	·0000	·5170
·20	·0000	·3263	·0000	·3835	·0000	·4892	·0000	·5993
·30	·0000	·4148	·0000	·4709	·0000	·5702	·0000	·6690
·40	·0552	·5367	·0229	·5720	·0000	·6438	·0000	·7297
0·50	0·1441	0·6400	0·0985	0·6741	0·0273	0·7289	0·0000	0·7838
·60	·2570	·7282	·2030	·7570	·1054	·8049	·0223	·8456
·70	·3942	·8068	·3379	·8290	·2259	·8657	·1059	·8985
·80	·5591	·8779	·5082	·8930	·3985	·9175	·2606	·9393
·90	·7578	·9422	·7235	·9498	·6436	·9620	·5271	·9727
0·95	0·8726	0·9719	0·8527	0·9757	0·8043	0·9818	0·7278	0·9871

See page 89 for explanation of the use of this table.

TABLE 31. BAYESIAN CONFIDENCE LIMITS FOR THE SQUARE OF A MULTIPLE CORRELATION COEFFICIENT

$k = 5$

CONFIDENCE LEVEL PER CENT

	90		95		99		99·9	
$n = 7$								
$R^2 = 0.00$	0·0000	0·4377	0·0000	0·5271	0·0000	0·6838	0·0000	0·8222
·10	·0000	·4563	·0000	·5459	·0000	·6998	·0000	·8327
·20	·0000	·4771	·0000	·5665	·0000	·7174	·0000	·8443
·30	·0000	·5011	·0000	·5895	·0000	·7361	·0000	·8566
·40	·0000	·5269	·0000	·6151	·0000	·7567	·0000	·8696
0·50	0·0000	0·5576	0·0000	0·6444	0·0000	0·7795	0·0000	0·8837
·60	·0000	·5940	·0000	·6783	·0000	·8052	·0000	·8991
·70	·0000	·6384	·0000	·7188	·0000	·8347	·0000	·9164
·80	·0000	·6955	·0000	·7695	·0000	·8700	·0000	·9362
·90	·0000	·7774	·0000	·8392	·0000	·9154	·0000	·9606
0·95	0·0509	0·8858	0·0000	0·8898	0·0000	0·9460	0·0000	0·9761
$n = 20$								
$R^2 = 0.00$	0·0000	0·1969	0·0000	0·2482	0·0000	0·3550	0·0000	0·4821
·10	·0000	·2525	·0000	·3115	·0000	·4276	·0000	·5557
·20	·0000	·3190	·0000	·3824	·0000	·5011	·0000	·6222
·30	·0000	·3943	·0000	·4584	·0000	·5718	·0000	·6830
·40	·0000	·4755	·0000	·5368	·0000	·6409	·0000	·7387
0·50	0·0445	0·5933	0·0079	0·6230	0·0000	0·7072	0·0000	0·7900
·60	·1433	·7074	·0879	·7378	·0104	·7791	·0000	·8374
·70	·2783	·7986	·2101	·8236	·0896	·8620	·0063	·8867
·80	·4557	·8758	·3864	·8931	·2432	·9199	·0897	·9416
·90	·6874	·9427	·6351	·9514	·5121	·9647	·3376	·9757
0·95	0·8310	0·9726	0·7986	0·9769	0·7165	0·9835	0·5824	0·9889
$n = 30$								
$R^2 = 0.00$	0·0000	0·1381	0·0000	0·1757	0·0000	0·2570	0·0000	0·3596
·10	·0000	·2092	·0000	·2576	·0000	·3543	·0000	·4649
·20	·0000	·2955	·0000	·3486	·0000	·4481	·0000	·5540
·30	·0000	·3879	·0000	·4408	·0000	·5354	·0000	·6314
·40	·0610	·5157	·0291	·5503	·0000	·6154	·0000	·6994
0·50	0·1556	0·6210	0·1115	0·6543	0·0391	0·7098	0·0000	0·7598
·60	·2735	·7119	·2230	·7399	·1288	·7876	·0399	·8310
·70	·4139	·7939	·3627	·8155	·2594	·8520	·1424	·8858
·80	·5788	·8687	·5339	·8835	·4370	·9079	·3124	·9304
·90	·7721	·9373	·7430	·9448	·6758	·9570	·5785	·9681
0·95	0·8812	0·9694	0·8647	0·9731	0·8251	0·9792	0·7642	0·9847

See page 89 for explanation of the use of this table.

TABLE 31. BAYESIAN CONFIDENCE LIMITS FOR THE SQUARE OF A MULTIPLE CORRELATION COEFFICIENT

$k = 6$

CONFIDENCE LEVEL PER CENT

	90		95		99		99·9	
$n = 8$								
$R^2 = 0·00$	0·0000	0·4005	0·0000	0·4861	0·0000	0·6406	0·0000	0·7845
·10	·0000	·4193	·0000	·5056	·0000	·6584	·0000	·7972
·20	·0000	·4404	·0000	·5271	·0000	·6776	·0000	·8114
·30	·0000	·4642	·0000	·5511	·0000	·6986	·0000	·8258
·40	·0000	·4916	·0000	·5783	·0000	·7216	·0000	·8415
0·50	0·0000	0·5234	0·0000	0·6094	0·0000	0·7473	0·0000	0·8586
·60	·0000	·5614	·0000	·6457	·0000	·7763	·0000	·8774
·70	·0000	·6083	·0000	·6896	·0000	·8100	·0000	·8984
·80	·0000	·6692	·0000	·7449	·0000	·8504	·0000	·9227
·90	·0000	·7575	·0000	·8217	·0000	·9027	·0000	·9524
0·95	0·0278	0·8520	0·0000	0·8778	0·0000	0·9381	0·0000	0·9713
$n = 20$								
$R^2 = 0·00$	0·0000	0·1969	0·0000	0·2482	0·0000	0·3550	0·0000	0·4821
·10	·0000	·2436	·0000	·3018	·0000	·4173	·0000	·5462
·20	·0000	·3013	·0000	·3644	·0000	·4835	·0000	·6080
·30	·0000	·3698	·0000	·4350	·0000	·5517	·0000	·6691
·40	·0000	·4477	·0000	·5114	·0000	·6206	·0000	·7238
0·50	0·0034	0·5358	0·0000	0·5914	0·0000	0·6887	0·0000	0·7769
·60	·0947	·6760	·0451	·7045	·0000	·7550	·0000	·8269
·70	·2257	·7821	·1568	·8079	·0469	·8444	·0000	·8737
·80	·4084	·8674	·3334	·8857	·1834	·9137	·0446	·9341
·90	·6556	·9393	·5967	·9486	·4584	·9627	·2671	·9742
0·95	0·8123	0·9711	0·7748	0·9757	0·6797	0·9827	0·5243	0·9884
$n = 30$								
$R^2 = 0·00$	0·0000	0·1381	0·0000	0·1757	0·0000	0·2570	0·0000	0·3596
·10	·0000	·1984	·0000	·2459	·0000	·3420	·0000	·4531
·20	·0000	·2764	·0000	·3297	·0000	·4305	·0000	·5389
·30	·0000	·3658	·0000	·4200	·0000	·5175	·0000	·6169
·40	·0327	·4839	·0047	·5151	·0000	·5998	·0000	·6871
0·50	0·1239	0·6022	0·0809	0·6357	0·0155	0·6881	0·0000	0·7506
·60	·2430	·6993	·1913	·7284	·0972	·7773	·0172	·8183
·70	·3872	·7855	·3339	·8080	·2269	·8459	·1094	·8806
·80	·5581	·8637	·5107	·8791	·4086	·9045	·2785	·9277
·90	·7600	·9350	·7289	·9428	·6569	·9556	·5525	·9670
0·95	0·8747	0·9683	0·8569	0·9722	0·8142	0·9786	0·7481	0·9843

See page 89 for explanation of the use of this table.

A NOTE ON INTERPOLATION

Part of the tabulation of a function $f(x)$ at intervals h of x is in the form given in the first two columns of the figure:

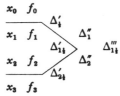

where $f_i = f(x_i)$ and $x_{i+1} = x_i + h$. Interpolation of $f(x)$ at values of x other than those tabulated uses the differences in the last three columns, where each entry is the value in the column immediately to the left and below minus the value to the left and above: thus, $\Delta'_{1\frac{1}{2}} = f_2 - f_1$ and $\Delta''_1 = \Delta'_{1\frac{1}{2}} - \Delta'_{\frac{1}{2}}$. These are usually written in units of the last place of decimals in $f(x)$. *Linear* interpolation between x_1 and x_2 approximates $f(x)$ by

$$f_1 + p\Delta'_{1\frac{1}{2}}$$

with $p = (x - x_1)/h$. This simple rule uses only the values within the lines of the figure and is often adequate. *Quadratic* interpolation between x_1 and x_2 approximates $f(x)$ by

$$f_1 + p\Delta'_{1\frac{1}{2}} - \tfrac{1}{4}p(1-p)\,(\Delta''_1 + \Delta''_2).$$

This is generally adequate provided $\Delta'''_{1\frac{1}{2}}$ is less than 60 in units of the last place of decimals in the tabulation. Notice that the quadratic interpolate consists of the addition of an extra term to the linear one, so that a rough assessment of it will indicate whether the linear form is adequate. The maximum possible value of $\tfrac{1}{4}p(1-p)$ is $\tfrac{1}{16}$ when $p = \tfrac{1}{2}$.

Example. The binomial distribution, $n = 20$, $r = 2$ (Table 1, page 22), interpolation in p, now x.

0·02	0·9929			
		−139		
·03	·9790		−90	
		−229		+3
·04	·9561		−87	
		−316		
·05	·9245			

For $x = 0·034$, $p = (0·034 - 0·03)/0·01 = 0·4$ and the linear interpolate is

$$0·9790 + 0·4 \times (-0·0229) = 0·9698.$$

The additional term for the quadratic interpolate is

$$-0·25 \times 0·4 \times 0·6 \times (-0·0090 - 0·0087) = 0·0011$$

and is not negligible, the quadratic interpolate being 0·9709. This is exact, as is expected since $\Delta'''_{1\frac{1}{2}}$ at 3 is well below 60.

The quadratic method uses f_0 and f_3 (needed for Δ''_1 and Δ''_2) and so fails if either is unavailable, for example at the ends of the range of x or when the interval of tabulation h changes. Modified quadratic forms between x_1 and x_2 are

$$f_1 + p\Delta'_{1\frac{1}{2}} - \tfrac{1}{2}p(1-p)\,\Delta''_1 \quad (f_3 \text{ missing}),$$
$$f_1 + p\Delta'_{1\frac{1}{2}} - \tfrac{1}{2}p(1-p)\,\Delta''_2 \quad (f_0 \text{ missing}).$$

Occasionally, *harmonic* interpolation is advisable. To do this the argument x is replaced by $1/x$ and then linear (or quadratic) interpolation performed.

Example. The F-distribution, $P = 10$, $\nu_1 = 1$ (Table 12(a), page 50), interpolation in ν_2, now x.

ν_2	$1/\nu_2$	$F(P)$		
∞	0	2·706		
			42	
120	1/120	2·748		1
			43	0
60	2/120	2·791		1
			44	
40	3/120	2·835		

Notice that the values of ν_2 chosen for tabulation are such that the intervals of $1/\nu_2$ are constant, here $\tfrac{1}{120}$. The differences show that linear interpolation will be adequate. For $\nu_2 = 80$, $p = (\tfrac{1}{80} - \tfrac{1}{120})/(\tfrac{1}{120}) = ·5$ and the linear interpolate is $2·748 + 0·5 \times 0·043 = 2·770$ with the possibility of an error of 1 in the last place.

CONSTANTS

$e = 2·71828\ 18285$	$\log_{10} e = 0·43429\ 44819$
$\pi = 3·14159\ 26536$	$\log_e 10 = 2·30258\ 50930$
$\dfrac{1}{\sqrt{2\pi}} = 0·39894\ 22804$	$\log_e \sqrt{2\pi} = 0·91893\ 85332$

Published by the Press Syndicate of the University of Cambridge
The Pitt Building, Trumpington Street, Cambridge CB2 1RP
40 West 20th Street, New York, NY 10011–4211, USA
10 Stamford Road, Oakleigh, Melbourne 3166, Australia

© Cambridge University Press 1984

First published 1984
Reprinted 1984, 1985, 1986, 1988 (with amendments), 1990, 1992, 1994
Second Edition 1995

A catalogue record for this book is available from the British Library

Library of Congress cataloguing in publication data available

ISBN 0 521 48485 5 paperback

Printed in Great Britain at the University Press, Cambridge